Analysis and Design of Hybrid Energy Storage Systems

Analysis and Design of Hybrid Energy Storage Systems

Special Issue Editor
Jorge Garcia

MDPI • Basel • Beijing • Wuhan • Barcelona • Belgrade

Special Issue Editor
Jorge Garcia
University of Oviedo
Spain

Editorial Office
MDPI
St. Alban-Anlage 66
4052 Basel, Switzerland

This is a reprint of articles from the Special Issue published online in the open access journal *Energies* (ISSN 1996-1073) from 2018 to 2020 (available at: https://www.mdpi.com/journal/energies/special_issues/Design_Hybrid_Energy_Storage).

For citation purposes, cite each article independently as indicated on the article page online and as indicated below:

LastName, A.A.; LastName, B.B.; LastName, C.C. Article Title. *Journal Name* **Year**, *Article Number*, Page Range.

ISBN 978-3-03928-686-7 (Pbk)
ISBN 978-3-03928-687-4 (PDF)

© 2020 by the authors. Articles in this book are Open Access and distributed under the Creative Commons Attribution (CC BY) license, which allows users to download, copy and build upon published articles, as long as the author and publisher are properly credited, which ensures maximum dissemination and a wider impact of our publications.

The book as a whole is distributed by MDPI under the terms and conditions of the Creative Commons license CC BY-NC-ND.

Contents

About the Special Issue Editor vii

Preface to "Analysis and Design of Hybrid Energy Storage Systems" ix

Ramy Georgious, Jorge Garcia, Pablo Garcia and Angel Navarro-Rodriguez
A Comparison of Non-Isolated High-Gain Three-Port Converters for Hybrid Energy Storage Systems
Reprinted from: *Energies* 2018, *11*, 658, doi:10.3390/en11030658 1

Muhammad Kashif Rafique, Zunaib Maqsood Haider, Khawaja Khalid Mehmood, Muhammad Saeed Uz Zaman, Muhammad Irfan, Saad Ullah Khan and Chul-Hwan Kim
Optimal Scheduling of Hybrid Energy Resources for a Smart Home
Reprinted from: *Energies* 2018, *11*, 3201, doi:10.3390/en11113201 25

Muhammad Umair Mutarraf, Yacine Terriche, Kamran Ali Khan Niazi, Juan C. Vasquez and Josep M. Guerrero
Energy Storage Systems for Shipboard Microgrids—A Review
Reprinted from: *Energies* 2018, *11*, 3492, doi:10.3390/en11123492 44

Pablo Quintana-Barcia, Tomislav Dragicevic, Jorge Garcia, Javier Ribas and Josep M. Guerrero
A Distributed Control Strategy for Islanded Single-Phase Microgrids with Hybrid Energy Storage Systems Based on Power Line Signaling
Reprinted from: *Energies* 2019, *12*, 85, doi:10.3390/en12010085 76

Pablo Arboleya, Islam El-Sayed and Bassam Mohamed
Modeling, Simulation and Analysis of On-Board Hybrid Energy Storage Systems for Railway Applications
Reprinted from: *Energies* 2019, *12*, 2199, doi:10.3390/en12112199 92

Macdonald Nko, S.P. Daniel Chowdhury and Olawale Popoola
Application Assessment of Pumped Storage and Lithium-Ion Batteries on Electricity Supply Grid
Reprinted from: *Energies* 2019, *12*, 2855, doi:10.3390/en12152855 113

Ramy Georgious, Jorge Garcia, Mark Sumner, Sarah Saeed and Pablo Garcia
Fault Ride-through Power Electronic Topologies for Hybrid Energy Storage Systems
Reprinted from: *Energies* 2020, *13*, 257, doi:10.3390/en13010257 149

About the Special Issue Editor

Jorge Garcia is Associate Professor at the Electrical Engineering Department at the University of Oviedo, in Spain, where he works with the LEMUR research team; his research interests include power electronic converters, including power topologies, control schemes and modeling, digital control, integration of stages, magnetic components, and industrial applications. Dr. Garcia is co-author of more than fifty international journal papers and more than one hundred international conference papers in power and industrial electronics. He was a visiting professor at the Federal University of Santa Maria, RS (Brazil), in 2012, at the University of Rome (La Sapienza), in 2013, at the University of Nottingham (UK) in 2016, and at the University of Aalborg (DK) in 2019.

Dr. Garcia is an associate editor of the International Journal in Electric Power & Energy Systems from Ed. Elsevier, and he has collaborated as a reviewer, guest editor and techincal chair in conferences and meetings of the most prestigious scientific publishers (IEEE, IET, MDPI).

Preface to "Analysis and Design of Hybrid Energy Storage Systems"

The most important environmental challenge today's society is facing is reducing the effects of CO_2 emissions and achieving a reduction of the effects of global warming in the environment. Such an ambitious challenge can only be achieved through a holistic approach, capable of tackling the problem from a multidisciplinary point of view.

A core technology that plays a critical role in this approach is the use of energy storage systems. These systems enable, among other things, the balancing of the stochastic behavior of renewable sources and distributed generation in modern energy systems; the efficient supply of industrial and consumer loads; the development of efficient and clean transport; and the development of nearly-zero energy buildings (nZEB) and intelligent cities.

Hybrid energy storage systems (HESS) consist of two (or more) storage devices with complementary key characteristics, that are able to behave jointly with better performance than any of the technologies considered individually. Recent developments in storage device technologies, interface systems, control and monitoring techniques, or visualization and information technologies have driven the implementation of HESS in many industrial, commercial, and domestic applications.

This Special Issue focuses on the analysis, design, and implementation of hybrid energy storage systems across a broad spectrum, encompassing different storage technologies (including electrochemical, capacitive, mechanical, or mechanical storage devices), engineering branches (power electronics and control strategies; energy engineering; energy engineering; chemistry; modeling, simulation and emulation techniques; data analysis and algorithms; social and economic analysis; intelligent and Internet-of-Things (IoT) systems; and so on.), applications (energy systems, renewable energy generation, industrial applications, transportation, uninterruptible power supplies (UPS) and critical load supply, etc.) and evaluation and performance (size and weight benefits, efficiency and power loss, economic analysis, environmental costs, etc.).

The Special Issue has made sound contributions to the state of the art of the topic in specific research lines, as it includes articles on the assessment of hybrid energy strategies in power systems ("Application Assessment of Pumped Storage and Lithium-Ion Batteries on Electricity Supply Grid", by M. Nko et al., and "Distributed Control Strategy for Islanded Single-Phase Microgrids with Hybrid Energy Storage Systems Based on Power Line Signaling ", by P. Quintana-Barcia et al.); on hybrid storage systems in transportation applications ("Energy Storage Systems for Shipboard Microgrids—A Review", by M.U. Mutarraf et al., and "Simulation and Analysis of On-Board Hybrid Energy Storage Systems for Railway Applications", by P. Arboleya et al.); on hybrid storage systems in smart homes ("Optimal Scheduling of Hybrid Energy Resources for a Smart Home", by M.K. Rafique et al.) and finally on power electronics topologies (with 2 contributions entitled "A Comparison of Non-Isolated High-Gain Three-Port Converters for Hybrid Energy Storage Systems" and "Fault Ride-Through Power Electronic Topologies for Hybrid Energy Storage Systems" from R. Georgious et al.).

I hope you will find these technical contributions useful.

Jorge Garcia
Special Issue Editor

Article

A Comparison of Non-Isolated High-Gain Three-Port Converters for Hybrid Energy Storage Systems

Ramy Georgious, Jorge Garcia *, Pablo Garcia and Angel Navarro-Rodriguez

LEMUR Group, Department of Electrical Engineering, University of Oviedo, ES33204 Gijon, Spain; eng.ramy.georgious@hotmail.com (R.G.); garciafpablo@uniovi.es (P.G.); navarroangel@uniovi.es (A.N.-R.)
* Correspondence: garciajorge@uniovi.es

Received: 14 February 2018; Accepted: 12 March 2018; Published: 15 March 2018

Abstract: This work carries out a comparison of non-isolated topologies for power electronic converters applied to Hybrid Energy Storage Systems. At the considered application, several options for three-port circuits are evaluated when interfacing a DC link with two distinct electrical energy storage units. This work demonstrates how the proposed structure, referred to as Series-Parallel Connection, performs as a simple, compact and reliable approach, based on a modification of the H-bridge configuration. The main advantage of this solution is that an effective large voltage gain at one of the ports is attained by means of a simple topology, preventing the use of multilevel or galvanic-isolated power stages. The resulting structure is thoroughly compared against the most significant direct alternatives. The analysis carried out on the switching and conduction losses in the power switches of the target solution states the design constraints at which this solution shows a performance improvement. The experimental validations carried out on a 10 kW prototype demonstrate the feasibility of the proposed scheme, stating its benefits as well as its main limitations. As a conclusion, the Series-Parallel Connection shows a better performance in terms of efficiency, reliability and controllability in the target application of compensating grid or load variations in Non-Isolated Hybrid Storage Systems, with large mismatch in the storage device voltage ratings.

Keywords: hybrid storage systems; power electronic converters; multiport; high gain converters; ultracapacitors

1. Introduction

At present, Hybrid Storage Systems (HSSs) are turning into one of the key technologies in power electronics related disciplines [1]. Indeed, by using these systems, there is a reported improvement in the performance at leading applications such as integration in the distribution network of stochastic power generators [2,3], grid stability and power quality support upon line contingencies [4], management of fast dynamics high power loads at the power-train in electric-hybrid vehicles [5,6], and a manifold of industrial applications with a load profile of large transient characteristics [7], among others. Generally speaking, these HSSs interface a fast-dynamics high-power storage device, e.g., a Ultracapacitor Module (UM), with a slower, bulk-energy storage unit, e.g., an Electrochemical Battery (EB) [8]. The design of the HSS involves the selection of adequate energy and power ratings in the elements of the system, as well as the design of a control scheme that manages properly the involved power flows [9]. The final design must ensure that the resulting HSS shows an overall enhanced performance, providing the energy ratings of the main energy storage device, but simultaneously maintaining the power ratings of the fast-dynamics one [1,10–12]. The management of the power flows in the system is generally implemented through Power Electronic Converters (PEC) that enable synchronized control and operation of the involved storage units [10–12].

Figure 1 depicts the power flow balance in a generic HSS. The primary energy source (in this case, it is the grid) supplies a given amount of power, P_{Grid}, to the front grid PEC (PECG). The aim of the system is to supply a power flow, P_{Load} towards another port that behaves either as a load or as a generator, for instance in the case of regenerative breaking applications. This port is interfaced through a Load/Generator PEC (PECL). At every instant, the difference between the load and grid power values, P_D, is managed by the control at the DC link. An adequate power balance into this DC link is essential for the correct operation of the system. A capacitor bank, DC link Cap in Figure 1, is usually employed as energy buffer for this difference power, being able to absorb or deliver the required P_{Cdc}. In some applications, for instance in islanded operation of microgrids, a very large energy storage capability is required at the DC link. The energy stored in the capacitor bank is normally not enough to ensure a stable operation at the DC link. Therefore, an extra energy storage system is interfaced to this DC link, ensuring also a fast recovery in case large power steps are demanded. Furthermore, in the case of hybrid systems, the total power of the HSS, P_{HSS}, is divided into two different storage units, ESS1 and ESS2, which are interfaced through two power converters, PEC ESS1 and PEC ESS2. The power through the EB and UM, P_{EB} and P_{UM}, respectively, is finally interfaced to the DC link.

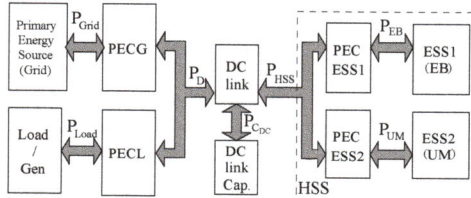

Figure 1. Scheme of the power flow balance in a general Hybrid Storage System (HSS).

To ensure a general case, all the power flows in the system must be considered as bidirectional. The conventional control scheme ensures that, upon normal conditions, i.e., power flowing from the grid to the load or from a generator back to the grid, the voltage at the DC link capacitor is kept constant. Then, in the event of sudden fluctuations at either load or line characteristics, the HSS control must compensate the resulting variation of operational parameters to ensure an adequate behavior of the system.

The simplest scheme for a three-port bidirectional converter interface in a HSS, able to attach these devices to a controlled DC link is the Direct Parallel Connection (DPC), can be seen in Figure 2a [13,14]. This scheme is formed by two independent bidirectional boost converters, each of them implemented by adding a filter inductor to a leg of a H-bridge converter. For the sake of clarity, a nominal DC link voltage of 600 V is defined in the coming discussion. In the same manner, the operating voltage ratings for the EB is considered to be in the range of 300–400 V (e.g., a Li-Ion EB intended for grid supporting applications). For these voltage ratios, a bidirectional boost converter can be selected as a feasible solution for interfacing both ports. Nevertheless, considering an extra storage unit with significant lower voltage ratings that requires to be interfaced, the voltage ratios between the DC link voltage and the storage unit voltage will change correspondingly. For instance, in the case of a UM as storage device of, e.g., 48 V voltage rating, and considering a steady-state reference value of 30 V, the direct interface through a boost topology would yield to the operation of the converter at duty ratios around 5%, well beyond the optimal 20–80% range [15]. On top of major concerns in the effect of parasitic elements, these extreme values for the duty ratio of the converter imply significantly high form factors in the voltage and current switch waveforms at the UM converter leg. In addition, it implies constraints in the practical control margins used in the regulation of the converter. All these issues, which are covered extensively in Section 2, make mandatory a search for simple, high-gain alternatives to interface low voltage ratings storage devices with the DC link [16–18].

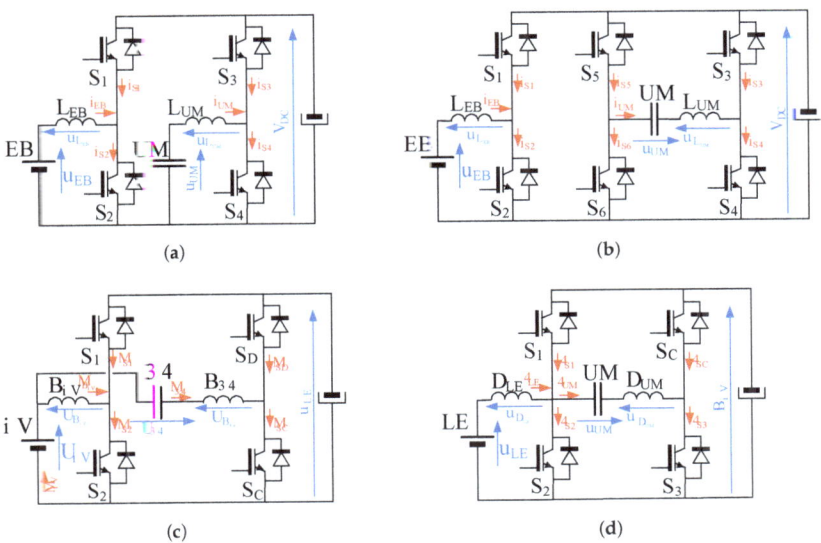

Figure 2. Non-isolated topologies considered for the HSS: (**a**) Direct Parallel Connection (DPC); (**b**) Full-Bridge Connection (FBC); (**c**) Series Connection (SC); and (**d**) Series-Parallel Connection (SPC).

The most used alternatives in high power applications include complex cascaded schemes (multistage solutions and tapped-inductor topologies), multilevel converters, or galvanically-isolated converters [19–25]. However, for small power levels, these solutions are not cost-competitive depending on the application. To solve these issues, the most straightforward solution in non-isolated converters is to use a Full Bridge Converter (FBC), as depicted in Figure 2b, for the UM storage system. As is justified in Section 3, this solution implies higher sizes and costs, as well as increased power losses. After that, Section 4 explores the non-isolated interfacing scheme proposed in [16], and depicted in Figure 2c, based on a Series Connection (SC) of the storage devices. After studying this scheme in detail, its advantages and drawbacks are highlighted. The most important drawback of this proposal is the series connection of both storage systems that eventually makes impossible an independent current control of both storage subsystems. This aspect will be discussed in Section 4. To overcome the mentioned drawbacks, an alternate solution is finally proposed for high voltage gain applications. The performance of the proposed Series-Parallel Connection (SPC) has been preliminarily explored on isolated applications [26]. However, this paper is focused on applying the SPC to the non-isolated scheme, as depicted in Figure 2d, and provides a deeper study than the one carried out in [27]. Therefore, this work aims to critically assess the performance of the SPC as a non-isolated alternative for HSS applications. This assessment is carried out through a detailed theoretical analysis, which is then validated by means of experimental performance demonstrations on a 10 kW rated laboratory setup. Section 5 covers the definition and detailed discussion of the switching modes in the converter, whereas Section 6 provides an analysis of the steady state operation of the topology. Section 7 evaluates the losses performance of the converter, and compares it to the DPC topology. From the conclusions derived of this study, Section 8 discusses the operation and suitability of this topology for HSS. A discussion on the control loop implemented for the validation is carried out in Section 9. Finally, Section 10 presents the final conclusions on the comparison carried out, and proposes some future work related with this topic.

2. Limitations of the Direct Parallel Connection

As mentioned, the DPC scheme in Figure 2a is suitable for the interconnection of DC sources when the voltage ratings at the storage systems are in the order of magnitude of half the DC link value. However, when a significantly small voltage ratings storage device is interfaced with a large DC link, i.e., with a ratio of 1:10 or higher, the bidirectional boost configuration is not the optimal option. As a large gain is required, the efficiency and the cost-effectiveness of the design are compromised [27,28]. Such a high gain requires a large duty ratio at the lower switch, over 90%. On the other hand, and considering a complementary pulses scheme, the remaining upper switch must be turned on with a very small duty ratio. These extreme duty ratio values yield to low efficiency [29]. As the switches must be designed for the high DC link voltage, high voltage ratings must be used. However, these devices present relatively large on-resistance values, and therefore conduction losses increase [30]. Moreover, in IGBT based topologies, this implies large currents at the antiparallel diodes, yielding to operation drawbacks derived from the reverse recovery phenomenon [31]. Moreover, the high duty cycles limit the switching frequency, as the minimum off-time of the switch must be ensured [32].

Finally, the dynamic performance of the converter is also affected, since the small duty ratios yield to non-symmetric bandwidths limitation in charge and discharge operation [27]. This last issue will be evidenced by considering an example of a HSS with the operating parameters of Table 2. In this case study, a 600 V DC link voltage is assumed, with a nominal operation voltages for the EB and the UM of 300 V and 30 V, respectively. All through this work, a particular notation will be used to clarify the discussion. Note that the subscript applied to the parameters for each of the studied topologies include a capital letter to distinguish the different configurations under consideration. In agreement with Figure 2, the magnitudes related to the DPC have a capital letter A in the subscript. In the same manner, the subscripts in the FBC include capital letter B. Letter C is used for subscripts in the SC, whereas subscripts for the SPC include capital letter D. For instance, the parameter D_1 (i.e., the duty ratio of Switch S_1) is represented as D_{1_A} for the DPC (Figure 2a), but is notated by D_{1_B} for the FBC (Figure 2b), and so on.

Initially, the steady state behavior is discussed. Upon these conditions, the corresponding duty ratios for the EB and UM legs are given by:

$$D_{1_A} = \frac{u_{EB}}{V_{DC}} = 50\% \tag{1}$$

$$D_{3_A} = \frac{u_{UM}}{V_{DC}} = 5\% \tag{2}$$

where D_{1_A} and D_{3_A} are the duty ratios of switches S_1 and S_3, respectively, for the DPC configuration. These relationships come directly from the gains of each leg of the bidirectional topology, which can be defined as:

$$M_{EB_A} = \frac{u_{EB}}{V_{DC}} \tag{3}$$

$$M_{UM_A} = \frac{u_{UM}}{V_{DC}} \tag{4}$$

where M_{EB_A} and M_{UM_A} are the static gain of the EB and the UM voltages to the DC link voltage, respectively, for the DPC configuration. In this case, the final capital letter A in the subscript indicates the DPC scheme.

Therefore, it is obvious to see that:

$$D_{1_A} = M_{EB_A} \tag{5}$$

$$D_{3_A} = M_{UM_A} \tag{6}$$

However, these expressions are a function of the topology, and will change for the rest of the topologies considered, as shown in the coming analysis.

2.1. Effects of the Waveform Shape in the Thermal Efforts

The final reliability of the design is a function of the relative value and distribution scheme of the thermal efforts associated to the electrical parameters [33]. The following paragraph discusses the effect of the shape of the waveforms in the distribution of the electrical stresses of a leg at the converter. Assuming small current ripples, then the form factor (K_f) of the current waveform for a given switch in the converter follows the general expression for a square waveform:

$$K_f = \frac{I_{rms}}{I_{avg}} \tag{7}$$

$$I_{rms} = I_{pk}\sqrt{\frac{T_{up}}{T_s}} \tag{8}$$

$$I_{avg} = I_{pk}\frac{T_{up}}{T_s} \tag{9}$$

$$K_f = \frac{1}{\sqrt{\frac{T_{up}}{T_s}}} \tag{10}$$

where I_{rms}, I_{avg} and I_{pk} are the rms, average and peak currents of a periodic square waveform of period T_s, respectively, being T_{up} the interval of the waveform that the current value equals I_{pk}.

In the DPC topology, for large mismatch between the ratings at the UM and at the DC link, the duty ratio at the UM leg (D_{3_A}) is close to 0%, around 5% in the case under study. This duty ratio at the upper leg can be expressed as:

$$D_{3_A} = \frac{T_{up}}{T_s} \tag{11}$$

and therefore:

$$K_{f_{3_A}} = \frac{1}{\sqrt{D_{3_A}}} \tag{12}$$

Analogously for the lower switch:

$$K_{f_{4_A}} = \frac{1}{\sqrt{1 - D_{4_A}}} \tag{13}$$

Therefore, for a duty ratio close to 5%:

$$K_{f_{3_A}} = 4.47 \tag{14}$$
$$K_{f_{4_A}} = 1.02 \tag{15}$$

This difference between the form factors at the switches of the leg of the UM converter implies that the thermal efforts at both switches are very different. Ideally, to evenly distribute these thermal efforts among the upper and lower switches of a leg, the duty ratios should be around 50%, yielding to K_f values close to:

$$K_{f_3} = K_{f_4} = 1.41 \tag{16}$$

2.2. Prototype and Experimental Setup

An experimental setup of the PC converter has been implemented in an existing laboratory prototype of 10 kW (Table 1). The prototype can be configured in all the configurations discussed in this work (Figure 3). The setup is built using a ROOK 48 × 6 module lithium-ion battery from CEGASA Portable Energy, a BMOD0165 P048 C01 Ultracapacitor Module from Maxwell Technologies and 2MB1200HH-120-50 IGBT modules from Fuji Electric, switching at a frequency of 20 kHz. The setup uses a TMS320F28335 from TI as control platform. The design uses standard reactive elements.

Table 1. Parameters of the 10 kW prototype.

Symbol	Parameter	Value
u_{EB}	Nominal Battery Voltage	288 V
u_{EB_min}	Minimum Battery Voltage (0% SOC)	225 V
u_{EB_max}	Maximum Battery Voltage (100% SOC)	328 V
i_{EB}	Battery Current	±30 A
u_{UM}	Rated UM Voltage	48 V
i_{UM}	UM Current	±200 A
C_{UM}	UM Capacitance	165 F

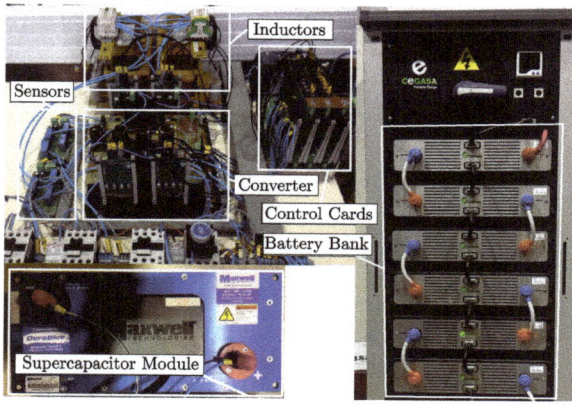

Figure 3. Experimental setup that can be configured as Parallel Connection (PC), Full Bridge Converter (FBC), Series Connection (SC) or Series-Parallel Connection (SPC).

For the conditions of the above description for the DPC scheme, with such a small D_{3_A}, the waveforms of the current through both switches S_3 and S_4 in Figure 2a present average values and K_f that are significantly different. This issue implies a high mismatch, both in the electrical and in the thermal stresses at each switch.

Figure 4a shows these waveforms at the switches, for the setup operating with the parameters in Table 2. As can be seen, the main currents and voltages measurements, consistent with the references in Figure 2a, are represented. It can be appreciated how the duty ratio of the switches reach extreme values. This yields to the aforementioned operational and design limitations, that eventually prevent the use of this topology.

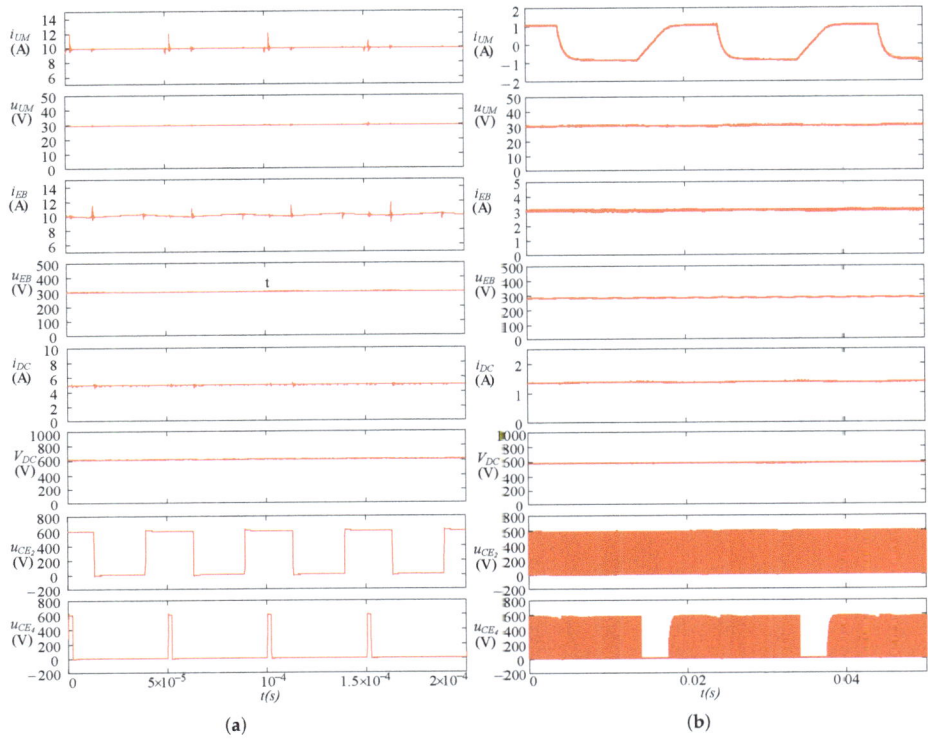

Figure 4. Experimental waveforms of the DPC scheme. Steady states: (**a**) i_{EB} = 10 A, i_{UM} = 10 A; and (**b**) i_{EB} = 10 A, i_{UM} steps from 1 A to −1 A.

Table 2. Operating parameters of the system under study.

Symbol	Parameter	Value
V_{DC}^*	DC link Voltage	600 V
R_{LOAD}	DC load resistor	300 Ω
u_{EB}	Battery Voltage	300 V
i_{EB}^*	Battery Current	10 A
u_{UM}	UM Voltage	30 V
i_{UM}^*	UM Current	10 A
f_s	Switching Frequency	20 kHz

As mentioned, the dynamic performance is also significantly affected by the extreme duty ratio values in the UM branch, which are very close to the 0–100% physical limits. Considering an abrupt negative step in the UM current demand, $i_{UM}*$, then the control stage must generate a control action in the duty ratios that provide the actual UM current, i_{UM}, equal to the reference value. However, the available control actions range from D_{3_A} = 5% to D_{3_A} = 0%, which ultimately implies S_3 and S_4 continuously turned off and on, respectively. This condition implies that the UM inductance, L_{UM}, is discharged with the relatively small voltage at the UM, u_{UM}, thus implying a limitation in the rate of decrease of the UM current. This aspect penalizes the discharging dynamics enormously, also introducing a non-symmetric behavior in the system performance. Indeed, for the opposite case (charging current), the extreme operation in the control action would imply a charging voltage of V_{DC}, and the rate of charge results dramatically increased.

This situation is illustrated in Figure 4b, where a series of alternate steps in the UM current reference, from 1 A to −1 A and vice-versa, are provided to the system. Even though these current values are several orders of magnitude smaller than the expected operational range, it can be seen how, at the beginning of the charging step (i.e., current i_{UM} changing from −1 A to 1 A), the modulation temporarily stops, as D_{3_A} results clamped to 0%. Therefore, the demanded control action would drop to negative values, yielding to an impossible operating constraint. On the other hand, for the discharging step (−1 A to 1 A), the control action can be provided by the system without constraints. Thus, the non-symmetrical performance of the system is demonstrated. Notice that this effect would take place even if the switches are considered ideal.

3. The Full Bridge Converter

To solve these issues, the most straightforward solution among non-isolated topologies is to use a FBC, as depicted in Figure 2b. During the following discussion, the inductors at the converter are considered ideal and purely inductive, therefore neglecting any parasitic resistances. This simplification is generally accurate for a reasonably good design of the magnetic devices. For the references at this figure, and considering an ideal inductor, the static gain of the UM leg of the converter can be defined as:

$$M_{UM_B} = \frac{u_{UM}}{V_{DC}} = \frac{D_{3_B} \cdot V_{DC} - D_{5_B} \cdot V_{DC}}{V_{DC}} = D_{3_B} - D_{5_B} \tag{17}$$

where D_{3_B} and D_{5_B} are the duty ratios of switches S_3 and S_5, respectively, for the FBC solution.

As can be seen, the effective gain between the UM and the DC link is the difference between the duty ratios of both legs of the H bridge converter. The value of the duty ratio of the UM branch can be calculated then as:

$$D_{3_B} = D_{5_B} + \frac{u_{UM}}{V_{DC}} = D_{5_B} + M_{UM_B} \tag{18}$$

It is assumed that the duty ratios of S_3 and S_5 are complementary to the ones at S_4 and S_6, respectively, as in the following scheme:

$$D_{4_B} = 1 - D_{3_B} \tag{19}$$
$$D_{6_B} = 1 - D_{5_B} \tag{20}$$

In this case, the effective static gain of the UM leg, M_{UM_B}, is a subtraction of both converter legs duty ratio levels. In other words, the voltage constraints impose the difference in the values of D_{3_B} and D_{5_B}, but the value itself can be selected arbitrarily. This implies that these duty ratios can reach more adequate values than in the boost converter case, while the difference can be made very small to achieve a large resulting gain. In fact, this effect comes as there is a new degree of freedom that can be selected to have one of the duty ratios, e.g., D_{5_B}, fixed and equal to 50%. This ensures effective duty ratios at each leg out away from the extreme values, i.e., within the 20–80% areas, therefore achieving better general performance [27,28,30–32]. In addition, the dynamic range is greater, given that the asymmetric modulation constraint of the DPC solution is not present any more. The payback in this case is the use of two additional switches in a second leg. This issue increases the size and weight of the converter, as well as the switching and conduction losses. However, with this solution the dynamics are not limited to the low duty ratios in the converter [27]. However, all four switches need to cope with the large voltages at the DC-link, V_{DC}, even though the device to interface presents significantly small ratings, yielding again to large conduction and switching losses [30,31]. A set of experiments has been carried out, configuring the same converter used in Figure 4 as a FBC. The system operates at a DC link voltage of 600 V, a UM voltage of 30 V, and different current values combinations flowing through both the EB and the UM. Table 3 shows experimental values of the losses and efficiency measurements of both DPC and FBC configurations, for the aforementioned voltage and current conditions, considering an extended set of UM current values (it must be noticed that the asterisks at any variable represent the references for the control systems). As can be seen, compared to the DPC operation, the FBC losses result increased, and therefore the efficiency of the FBC configuration is significantly smaller.

Table 3. Experimental losses and efficiency performance of DPC and FBC configurations.

i_{EB}^* (A)	i_{UM}^* (A)	$P_{Loss}(DPC)$ (W)	$P_{Loss}(FBC)$ (W)	η_{DPC} (%)	η_{FBC} (%)
10	10	308.6	434.3	90.6%	86.6%
10	5	251.8	339.4	92.0%	89.3%
10	0	222.7	250.4	92.6%	91.8%
10	−5	247.5	310.8	91.7%	89.4%
10	−10	286.4	374.7	90.4%	86.7%

4. Series Connection of the Storage Systems

Figure 2c shows the series configuration of the storage systems [10]. It must be noticed how the UM is connected in series to the EB. The analysis of the topology starts by looking at the mesh equation that relates the voltages at the switches S_2 and S_4, at both inductors and at the UM, with the references in Figure 2c:

$$u_{CE_2}(t) + u_{UM}(t) + u_{L_{EB}}(t) - u_{L_{UM}}(t) = u_{CE_4}(t) \tag{21}$$

where $u_{CE_2}(t)$ and $u_{CE_4}(t)$ are the collector to emitter voltages of switches S_2 and S_4, respectively, and $u_{L_{EB}}(t)$ and $u_{L_{UM}}(t)$ are the voltages at the inductors L_{EB} and L_{UM}, in the EB and UM legs, respectively. The average inductor voltages will be null at steady state, and thus Equation (21) can be expressed as:

$$u_{CE_2} + u_{UM} = u_{CE_4} \tag{22}$$

Given that each leg of the H-bridge operates as a bidirectional boost converter, the average values of $u_{CE_2}(t)$ and $u_{CE_4}(t)$ are again a function of the duty ratios at the upper switches of the H-bridge converter, D_{1_C}, for S_1 at the EB leg, and D_{3_C}, for S_3 at the UM leg, respectively. The expressions for the static gain in SC is analog to the ones derived in Equations (3) and (4), for FBC, but for consistency, they are expressed for this topology as:

$$M_{EB_C} = \frac{u_{EB}}{V_{DC}} \tag{23}$$

$$M_{UM_C} = \frac{u_{UM}}{V_{DC}} \tag{24}$$

Equations (23) and (24) yield to the expression for the duty ratios, D_{1_C} and D_{3_C}, and the static gains, M_{UM_C} and M_{EB_C}, in steady state:

$$D_{1_C} = \frac{u_{EB}}{V_{DC}} = M_{EB_C} \tag{25}$$

$$D_{3_C} = \frac{u_{UM} + u_{EB}}{V_{DC}} = M_{UM_C} + M_{EB_C} \tag{26}$$

From Equations (25) and (26), the expression that relates the duty ratio from both legs can be calculated as:

$$D_{3_C} = D_{1_C} + \frac{u_{UM}}{V_{DC}} = \frac{u_{UM} + u_{EB}}{V_{DC}} \tag{27}$$

Equation (27) is interesting, since the duty ratio of switch S_3 at the UM leg, i.e., D_{3_C}, is not a function of the UM and DC voltage values alone (which would yield to very small duty ratio values as in the DPC), but also a function of the battery voltage. This is has a similar effect to what was found for the FBC scheme. From Equation (27), for the operating conditions in Table 2, the duty ratio values at the UM leg of the converter for the DPC change from D_{3_A} = 5% for the DPC to D_{3_C} = 55% in the SC. Therefore, again, the resulting duty ratios at the SC scheme imply a significant improvement in the current stresses balancing versus the DPC case. In addition, the control margins for the control actions achievable in this topology result increased, and in principle this would allow for a symmetrical fast dynamics performance design, in line with the FBC case. The only constraint in the design for SPC scheme is that the voltage at the battery as well as the voltage at the UM cannot reach the DC link voltage. This means the duty ratio of the switch S_3 at the UM leg must be less than or equal 100% ($D_{3_C} \leq 100\%$). However, for practical values, a feasible system ensuring this condition can be designed without major issues.

However, the most significant drawback in this scheme comes from the expression of the battery current, which can be expressed by:

$$i_{EB} = i_{L_{EB}} + i_{UM} \qquad (28)$$

where i_{EB}, $i_{L_{EB}}$ and i_{UM} are the currents of the EB, inductor L_{EB} and UM, respectively.

This results in the impossibility of implementing a practical decoupled current control scheme in both storage systems (EB and UM). In fact, if both inductor currents are independently controlled, then the evolution of the battery inductor current is forced by Equation (28), yielding either to dangerous voltages in the system due the inductive behaviour, or to a limited dynamic performance if these overvoltages are prevented at control level.

Another point of the analysis comes by looking in Figure 2c. From the inductors connection scheme, it might seem that a certain beneficial interleaving effect is possible in the EB current, i_{EB}. Nevertheless, this effect would only be true for small operating conditions ranges, as it depends on the values of the duty ratios and in the synchronization of the pulses in the switches. This enhanced interleaving effect will not occur for all possible conditions, particularly for UM currents much higher than EB currents.

Finally, also derived from Equation (28), the peak current flowing through the EB inductor is calculated as a function of the UM and EB currents. Thus, inductor L_{EB} must be designed considering values in the order of magnitude of the UM current, that is, significantly larger than the EB current. This results in a much larger inductor device, which compromises the efficiency, the power density and the cost of the full HSS.

Given all these constraints, the SC scheme is disregarded as a feasible option. Therefore, it will not be included for the validation stages, by simulations or experimental tests.

5. Analysis of the Series-Parallel Connection

All these drawbacks of the SC scheme can be effectively solved by considering the SPC of both storage units. This scheme, shown in Figure 2d, keeps the H-bridge configuration of the switches. However, in this case, the series assembly formed by the EB and inductor L_{EB}, is connected between both midpoints of the legs. This configuration can be seen as an integration of the FBC from three to two legs, removing the degree of freedom that existed in the latter. For the references in Figure 2d, the mesh equation that includes the voltage at the UM can be expressed as:

$$u_{CE_2}(t) + u_{UM}(t) - u_{L_{UM}}(t) = u_{CE_4}(t) \qquad (29)$$

Analyzing Equation (29) analogously to the former FBC and SC cases, the expression for the duty ratio at the UM leg, results in:

$$D_{3_D} = M_{UM_D} + M_{EB_D} = \frac{u_{UM} + u_{EB}}{V_{DC}} = D_{1_D} + \frac{u_{UM}}{V_{DC}} \qquad (30)$$

Again, it presents a similar expression to the SC case; thus, all statements concluded for the new duty ratio values are still valid.

Regarding the stresses distribution, the values of the K_f can be calculated for the switches at the SM branch at the SPC configuration, considering the values of Table 2:

$$K_{f_{3D}} = \frac{1}{\sqrt{0.55}} = 1.35 \tag{31}$$

$$K_{f_{4D}} = \frac{1}{\sqrt{0.45}} = 1.49 \tag{32}$$

These values are close to the optimal value for an even distribution of the current efforts stated in Equation (16), therefore increasing the reliability of the system. This effect is obtained for any application in which one of the legs at the converter interfaces a device with voltage ratings significantly smaller than the other one, this latter being around half (e.g., practical values of 40–60%) the DC link value.

However, in addition to that, it must be noticed how, in the SPC scheme, the following expression can be calculated for the EM current:

$$i_{EB} = i_{L_{EB}} \tag{33}$$

Therefore, and unlike in the SC case, decoupling both EB and UM current control is quite simple in the SPC scheme. This results in the possibility of implementing an independent current control (and hence power flow) for both storage devices. This allows for an effective hybridization of the energy devices, without the drawbacks of extreme duty rations in the system.

The following discussion deals with a deep analysis of the operation of the SPC converter, aiming to provide the foundations for an adequate design of the HSS. The previous step of the analysis is to settle the assumptions and limitations that are going to be considered to simplify and establish the limits of the study. These assumptions are the following:

- The UM is a unipolar DC device, and the terminal of negative polarity is attached to the center point of the battery leg. Thus, it can be deduced from (30) that D_{3_D} is greater than D_{1_D} in steady state.
- The switching pulses of all the switches are synchronized at the same frequency, f_S.
- The ripple values of the current through both inductors and of the voltage at the capacitor, are relatively smaller than the respective average values.
- Each leg at the converter operates in a complementary scheme, i.e., the pulse signals for the lower switches are the logical inverted pulses of the upper ones. It is also assumed that a dead time is implemented in the switching scheme, aiming to avoid cross-conduction, and that its effect in the overall performance can be neglected.
- The initial conditions assume a positive value for i_{EB}, i.e., the battery is being discharged towards the DC link.
- Finally, it is also considered a positive value for i_{UM}, i.e., the UM is also being discharged. However, to increase the generality of the analysis, in a later stage, the case of negative i_{UM} will also be considered.

Once the basic operating assumptions are settled, the instant waveforms at the converter must be analyzed. However, the shapes of these waveforms depend on the exact sequence of gating signals in the switches. Each leg operates in complementary mode, as stated previously; however, in the most general case, the phase shift between legs might take any value, resulting in different synchronization schemes. From the point of view of the implementation of the PWM scheme in a digital controller, the most straightforward manner to synchronize the pulses is to use a single triangular waveform at f_S, and compare this triangular shape with given reference values to generate the control pulses for every switch in the converter. For this single triangular waveform scheme, the pulses obtained are symmetrical from the central point of the on/off intervals, as depicted in Figure 5. In particular,

Figure 5a corresponds to the switching pattern in the steady state, i.e., for D_{3_D} being greater than D_{1_D}. This gives rise to a set of operating modes, as a function of the combination of on/off states of the switches in the converter. These equivalent switching modes are detailed in the following subsections, considering the chronograms in Figure 5.

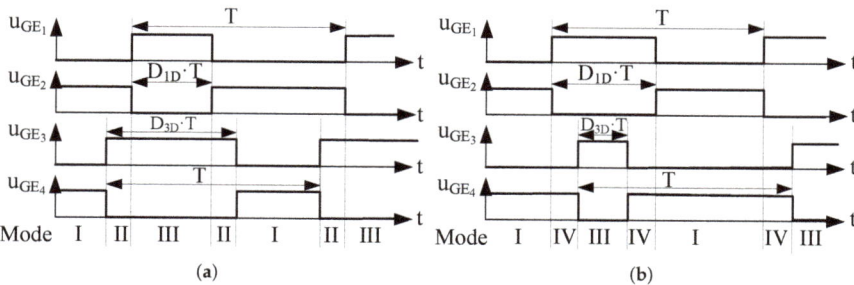

Figure 5. Switching modes in the SPC for the pulse scheme considered: (**a**) D_{3_D} is greater than D_{1_D}; and (**b**) D_{3_D} is smaller than D_{1_D} (only in transients).

5.1. Mode I. S_2 and S_4 Turned On

Figure 6a shows both S_2 and S_4 turned on. The battery inductor charges through S_2 ($i_{EB} > 0$). Assuming also $i_{UM} > 0$, then L_{UM} charges through S_2 and S_4:

$$\begin{aligned} i_{S1}(Mode\ I) = 0; \quad & i_{S3}(Mode\ I) = 0; \\ i_{S2}(Mode\ I) = i_{EB} - i_{UM}; \quad & i_{S4}(Mode\ I) = i_{UM}; \end{aligned} \tag{34}$$

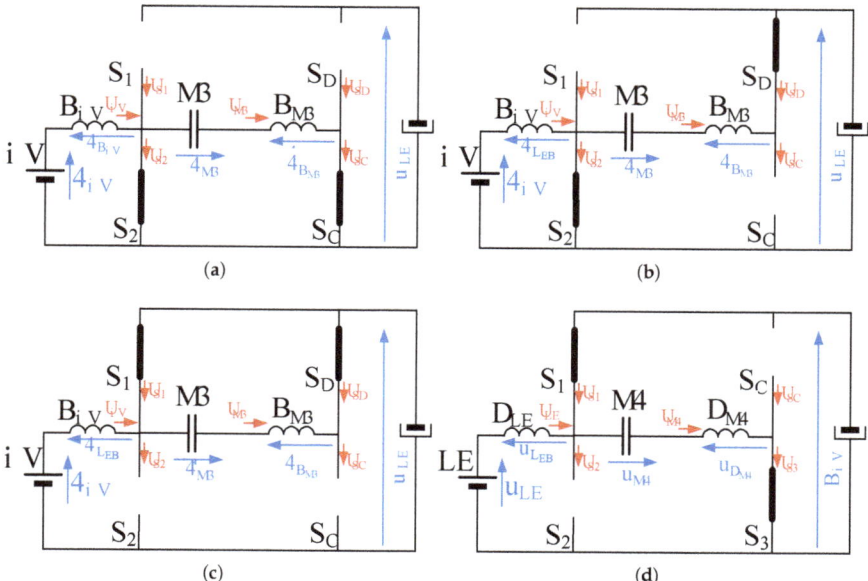

Figure 6. Switching modes in the SPC scheme: (**a**) Mode I; (**b**) Mode II; (**c**) Mode III; and (**d**) Mode IV.

5.2. Mode II. S_2 and S_3 Turned On

In the next switching interval, depicted in Figure 6b, S_4 turns off and S_3 turns on, whereas the battery leg remains unchanged. The UM current flows towards the DC link through S_3, and therefore:

$$i_{S1}(Mode\ II) = 0; \quad i_{S3}(Mode\ II) = -i_{UM}; \\ i_{S2}(Mode\ II) = i_{EB} - i_{UM}; \quad i_{S4}(Mode\ II) = 0; \tag{35}$$

5.3. Mode III. S_1 and S_3 Turned On

Finally, mode III keeps the UM leg as in Mode II, but now S_1 is turned on as S_2 turns off (Figure 6c). The resulting current expressions in the switches for this interval are:

$$i_{S1}(Mode\ III) = -i_{EB} + i_{UM}; \quad i_{S3}(Mode\ III) = -i_{UM}; \\ i_{S2}(Mode\ III) = 0; \quad i_{S4}(Mode\ III) = 0; \tag{36}$$

5.4. Mode IV. S_1 and S_4 Turned On

An additional switching mode has to be analyzed. During transients, D_{3D} might get smaller than D_{1D}, and therefore Mode IV would take place instead of Mode II (see Figure 6b) in the switching sequence. In this case, S_1 and S_4 will be turned on, whereas S_2 and S_3 will remain turned off (Figure 6d):

$$i_{S1}(Mode\ IV) = -i_{EB} + i_{UM}; \quad i_{S3}(Mode\ IV) = 0; \\ i_{S2}(Mode\ IV) = 0; \quad i_{S4}(Mode\ IV) = i_{UM}; \tag{37}$$

6. SPC Steady State Analysis

Once the switching states are defined, the steady state analysis of the SPC scheme can be carried out. It must be noticed that both converters are bidirectional in current, and thus, if a general analysis is desired, all possible combinations must be assessed. Considering a system that operates with DC link voltage control, and provided that both storage device legs are controlled in current mode, the operating conditions that need to be taken into account are stated in Table 4.

Table 4. Operating conditions of storage systems, considering references in Figure 6.

EB	UM	Operating Condition
Discharging $i_{EB} > 0$	Charging $i_{UM} < 0$	Opposite sign in currents
Discharging $i_{EB} > 0$	Discharging $i_{UM} > 0$	Same sign in currents
Charging $i_{EB} < 0$	Charging $i_{UM} < 0$	Same sign in currents
Charging $i_{EB} < 0$	Discharging $i_{UM} > 0$	Opposite sign in currents

From Equations (34)–(37), the current that flows through the switches at the battery leg are a subtraction of the EB and UM inductor currents. Therefore, the net result of these switch currents depends on whether these currents are added or subtracted in absolute value. Thus, this study can be simplified to the cases in which UM and EB currents have either the same or opposite signs. The theoretical waveforms for these two key cases can be seen in Figure 7a (EB and UM discharging, i_{EB} and i_{UM} have same signs) and Figure 7b (EB discharging, UM charging, i_{EB} and i_{UM} present opposite signs). Even if the resulting current values at the switches result in significant change, the claimed balancing effect in the current stresses at the UM leg switches can still be noticed, as all the involved duty ratios are relatively close to the 50% optimal value.

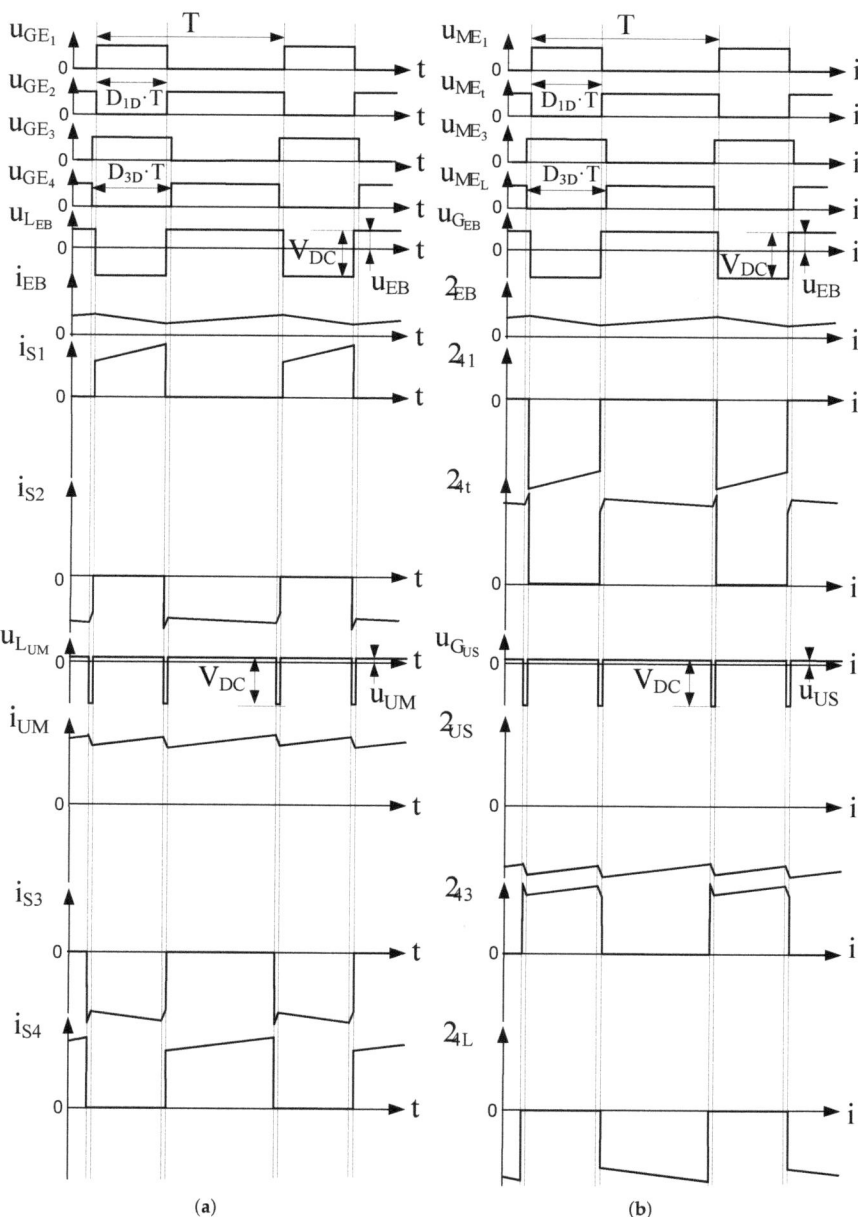

Figure 7. Theoretical waveforms of the SPC scheme: (**a**) EB and UM both are in discharge mode; and (**b**) EB is in discharge mode but UM is in charge mode.

Another consequence of this switching pattern is that the current waveforms through the UM inductor evolves at twice the switching frequency. This allows for a certain degree of optimization in the inductor design, as current ripple will decrease for the same target inductor value, or, conversely, inductor can be made smaller for the same target current ripple. Figure 8a shows key experimental waveforms measured at steady state, for the SPC configuration of the prototype setup defined in Table 2.

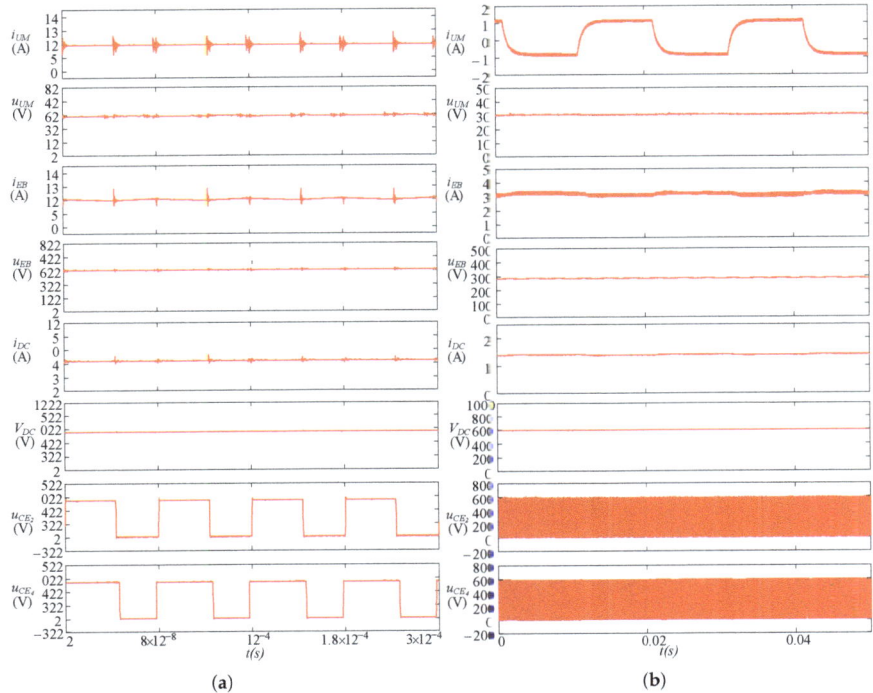

Figure 8. Experimental waveforms of the SPC scheme. Steady states: (**a**) i_{EB} = 10 A, i_{UM} = 10 A; and (**b**) i_{EB} = 3 A, i_{UM} steps from 1 A to −1 A.

However, the most significant consequence of this connection comes from the relationship between M_{EB_D} and M_{UM_D}, and therefore between D_{1_D} and D_{3_F}. As stated in the assumptions, and considering the steady state operation, then from Equation (30), D_{3_D} is always greater than D_{1_D}. However, in transient operation, the inductor voltage at the UM might be substantially large, depending on the transient current demanded. This might yield D_{3_D} to reach values smaller than D_{1_D}. However, as in the FBC case, now the control action at the UM is not clamped as in DPC, and therefore a better dynamic performance is found. Moreover, this behavior is now symmetric. To illustrate this last assertion, Figure 8b shows a series of symmetric ±1 A consecutive bidirectional current steps for the SPC connection, in order to keep the same values as in the DPC case (Figure 4b), as to be compared directly. As it can be seen, and given that the control action is able to reach negative values in a natural manner, the modulation is never interrupted in the SPC operation. Figure 9 shows the performance for an increased current steps than in Figure 8b, up to ±10 A current steps. As it can be seen in this figure, a large ripple an be appreciated in the DC link voltage. This is due the fact that for these experiments, the DC link is not regulated with optimal bandwidth. However, the aim of these plots is to show how this topology can supply large, fast current steps to the DC link voltage by the UM storage subsystem.

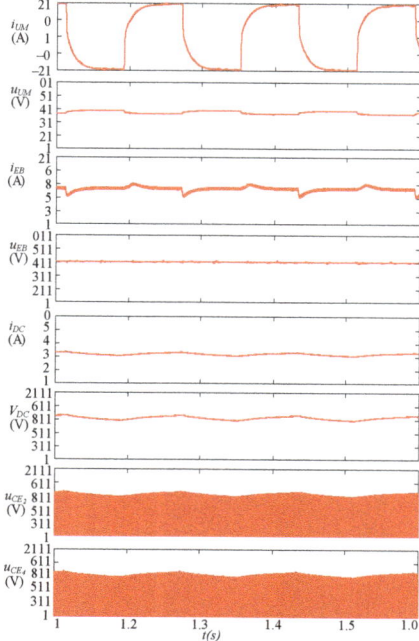

Figure 9. Experimental waveforms of the SPC scheme, for i_{UM} steps from 10 A to −10 A.

7. Losses Comparison and Effects in the Efficiency

From the above discussion, the SPC can be initially considered as an alternative solution for a non-isolated interface in a HSS, in the case that one of the storage devices is rated at very low voltage. As can be seen, the proposed scheme overcomes the main drawbacks of the DPC, FBC and SC schemes. However, major concerns in the performance of the solution arise from the fact that the UM current will flow also through the switches of the EB leg. It means that both switching and conduction losses through these switches will be affected. In the event that the final losses at these switches result in an increase with respect to the original scheme, the overall efficiency loss might make unfeasible the use of this solution. Moreover, as is demonstrated below, the final balance depends on the operation point of the HSS. Thus, a thorough, objective analysis of the time evolution of losses in the system as a function on the mode of operation must be carried out.

To assess this comparison quantitatively, the losses at every switch of converter have been expressed following a simplified theoretical approach. The generic equations of both the switching and conduction losses, for inductive switching of the converter, have been expressed as a function of the EB and UM current values [15]. However, to extract conclusions on the comparison of performances, the figure of merit that is considered is the difference between the losses at both the DPC and SPC configurations, ΔP_{Loss}, rather than the losses at each of the schemes on their own. Thus:

$$\Delta P_{Loss} = P_{Loss}(SPC) - P_{Loss}(DPC) \tag{38}$$

This parameter has been quantified theoretically for the operating parameters in Table 2, and the results are shown in Figure 10. This picture represents ΔP_{Loss} in a grey scale. The darker areas correspond to larger negative differences, i.e., the proposed SPC performs with fewer losses than the original DPC. Conversely, the clearer regions imply larger positive differences, i.e., SPC performs with more losses than DPC. As a conclusion, a better efficiency is obtained by using the proposed SPC scheme if the system evolves within at the darker areas. This implies both UM and EB currents are large in amplitude and of the same sign, that is to say, both storage devices are simultaneously being either charged or discharged.

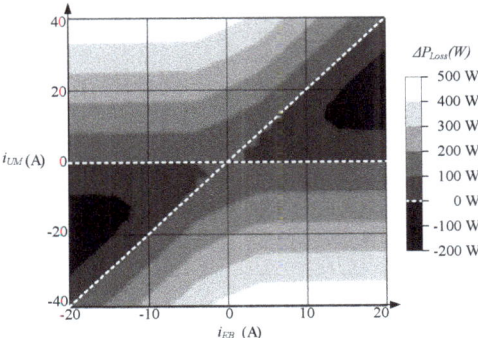

Figure 10. Difference between the losses in DPC and SPC configurations, as a function of the UM and EB currents. The darker areas correspond to SPC scheme operating with fewer losses than the original DPC scheme. For reference, it must be noted that P_{Loss} equals to zero if $i_{UM} = 0$ (i.e., horizontal axis).

The conclusions of this analysis have been validated through a series of experiments on the built setup. Table 5 shows efficiencies and losses obtained in steady state, of both the DPC and SPC configurations, for the known given voltage conditions of $V_{DC} = 600$ V, $u_{EB} = 300$ V and $u_{UM} = 30$ V. The recorded current reference values considered were $i_{UM} = -10$ A, -5 A, 0 A, $+5$ A and $+10$ A, and $i_{EB} = 0$ A, $+5$ A, $+10$ A. From these results, it can be verified that the switches losses are lower in SPC provided that the UM and EB currents are both large and of the same sign. On the other hand, if the signs of both currents are opposite, SPC presents more losses than DPC. Thus, Table 5 corroborate the theoretical results depicted in Figure 10. The results in Table 5 are graphically represented in Figure 11. As can be seen, the SPC presents fewer losses when the EB and UM currents have the same sign and larger values. As a conclusion, it must be noticed that, even though at some operating points the losses will be higher with the proposed SPC scheme than in the original DPC scheme, the full performance in terms of efficiency of the proposed topology must be assessed only after considering the application and the control scheme used.

Table 5. Experimental losses and efficiency performance of SPC and DPC configurations

i_{EB}^* (A)	i_{UM}^* (A)	$P_{Loss}(DPC)$ (W)	$P_{Loss}(SPC)$ (W)	ΔP_{Loss} (W)	η_{DPC} (%)	η_{SPC} (%)
0	10	142.0	248.6	106.6	-	-
0	5	82.0	147.9	65.9	-	-
0	0	20.3	15.1	−5.3	-	-
0	−5	99.8	55.3	−44.5	-	-
0	−10	206.0	250.8	44.8	-	-
5	10	240.1	224.4	−15.7	87.0%	87.4%
5	5	177.8	139.0	−38.8	89.5%	91.5%
5	0	146.0	138.3	−7.7	90.5%	91.0%
5	−5	166.0	191.8	25.9	89.1%	87.4%
5	−10	188.1	277.5	89.4	87.5%	81.9%
10	10	308.6	257.2	−51.5	90.6%	92.1%
10	5	251.8	218.7	−33.1	92.0%	93.1%
10	0	222.7	215.9	−6.9	92.6%	92.9%
10	−5	247.5	293.9	46.4	91.7%	90.3%
10	−10	286.4	363.9	77.5	90.4%	88.0%

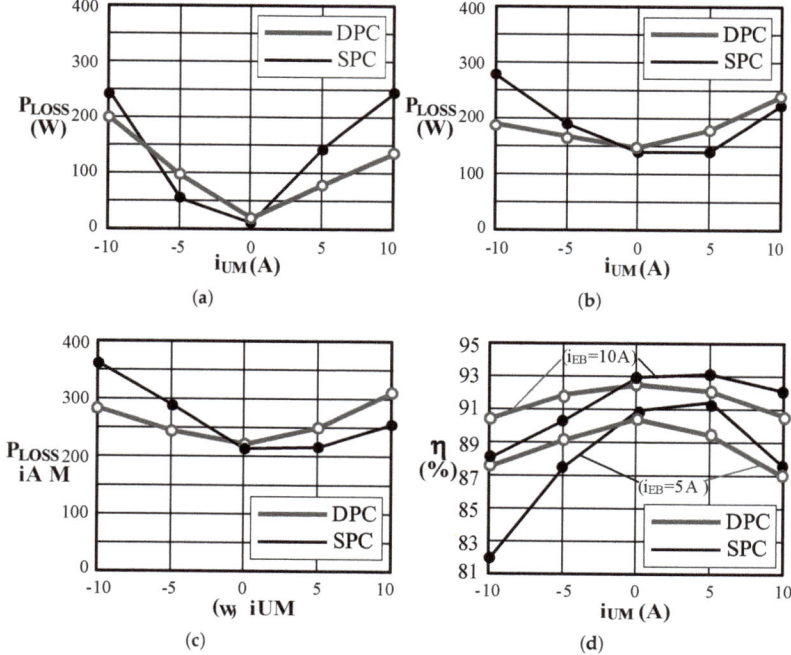

Figure 11. Steady state losses comparison between DPC and SPC configurations: (**a**) $i_{EB} = 0$ A; (**b**) $i_{EB} = 5$ A; and (**c**) $i_{EB} = 10$ A. (**d**) Efficiency measurements of the DPC and SPC schemes, for $i_{EB} = 5$ A and $i_{EB} = 10$ A.

8. SPC Scheme in Hybrid Storage Systems Applications

Thus far, the comparison of losses has been carried out considering steady state conditions. This section, instead, deals with the analysis of the SPC performance in HSS applications upon transient operation. For the power flows stated in Figure 1:

$$P_{C_{DC}} = P_{Grid} - P_{Load} + P_{EB} + P_{UM} \tag{39}$$

$$P_D = P_{Grid} - P_{Load} \tag{40}$$

$$P_{ESS} = P_{EB} + P_{UM} \tag{41}$$

where P_{Cdc} is the power absorbed by the DC link capacitor, P_{Grid} is the power coming from the grid, P_{Load} is the power consumed by the load, and P_{EB} and P_{UM} are the power flowing from both the EB and UM, respectively, towards the DC link. These power values are defined as a function of the voltage and current values at each subsystem [14]:

$$P_{Grid} = V_{DC} \cdot i_{Grid_{DC}} \tag{42}$$

$$P_{Load} = V_{DC} \cdot i_{DC} \tag{43}$$

$$P_{EB} = u_{EB} \cdot i_{EB} = V_{DC} \cdot i_{EB_{DC}} \tag{44}$$

$$P_{UM} = u_{UM} \cdot i_{UM} = V_{DC} \cdot i_{UM_{DC}} \tag{45}$$

$$P_{C_{DC}} = V_{DC} \cdot i_{C_{DC}} \tag{46}$$

where $i_{Grid_{DC}}$, i_{DC}, $i_{EB_{DC}}$, $i_{UM_{DC}}$ and $i_{C_{DC}}$ are the currents of the grid, the load, the EB, the UM and the DC link capacitor, respectively, all of them at the DC link side. In the system under consideration, the DC link voltage is regulated and fixed to a reference value. Therefore, in steady state F_{Cdc} is null. Assuming that all the load power is supplied by the grid converter, then the storage system remains idle in steady state, which means that also P_{EB} and P_{UM} are null. Thus, from Equation (39), the following equality applies in steady state:

$$P_{Grid} = P_{Load} \qquad (47)$$

However, upon transient variations modeled by power steps in either the grid (line fluctuations) or in the load (random load/stochastic generator), the balance given by Equation (47) is lost. It yields to a transient change in the DC link steady voltage value, that must be compensated by the control scheme if a stable operation is desired [34]. For simplicity, it is assumed that, once the system is in steady state, a load instant power step takes place at a given moment (i.e., the grid power is kept constant). It is also assumed that the hybrid behavior is designed as to achieve UM dynamics (power support) much faster than the EB dynamics (energy support) [35].

Figure 12 sketches this evolution. Interval 1 shows the initial steady state situation, when no power is flowing from any of the storage systems to the DC link. The power that the load is consuming is fully delivered by the grid. The storage system is in idle mode, and thus the currents flowing through the storage devices are null. Interval 2 starts with a sudden load change, in this case a step increase in the load. The control stage reacts demanding more power from the HSS. Therefore, both storage devices start to supply energy to the system. Due to the system constraints, the battery has limited safe dynamic response, and hence the power is initially supplied by the UM converter. In any case, P_{UM} and P_{EB}, and therefore i_{EB} and i_{UM}, present the same polarity. As per the aforementioned discussion, this results in smaller current stresses in the switches at the EB leg. Notice that an analogous situation is achieved in the case of decreasing step in the load power.

Once the energy supply is taken over by the EB, the UM must recharge to reach the initial reference value again in a reasonable amount of time. This ensures the HSS is ready to supply again any forthcoming power steps. However, this implies that the sign of the UM current changes, yielding to Interval 3. In this situation, EB and UM currents present opposite signs, resulting in an increase of the stresses at the switches of the EB leg. Nevertheless, this evolution back to idle mode might be done relatively slowly, allowing to minimize the effect of the addition of currents. Then, provided that the control dynamics are tuned adequately, SPC provides a better efficient performance than DFC.

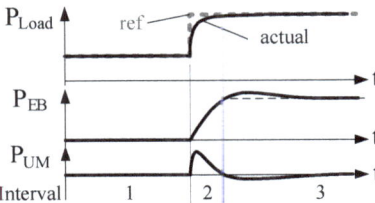

Figure 12. Control operation intervals upon sudden load variation.

9. Stability of the SPC Scheme

Figure 13a shows the scheme of the UM current mode control loop. Although the standard EB current mode and DC link voltage mode control schemes are not represented, it is assumed that these loops are operating properly. $H(s)$ is a signal conditioning block, in charge of measuring, adapting and filtering the current through the UM. The obtained measured value, $I_{UM_{Meas}}$, is compared to the reference, $I_{UM}{}^*$, to obtain the UM current error, ϵ_I. This error is the input of the regulator $R(s)$. The output of this regulator is the control action that enters the transfer function of the system, $G(s)$.

As seen in Figure 2d, the UM storage device current equals the UM inductor current, and hence the UM control is indeed an inductor current control. Such a control scheme can be implemented considering the inductor voltage, $U_{L_{UM}}$, as the control action. This yields to a transfer function given by:

$$G(s) = \frac{I_{L_{UM}}}{U_{L_{UM}}} = \frac{1}{s \cdot L_{UM} + R_{L_{UM}}} \tag{48}$$

where $R_{L_{UM}}$ is the parasitic resistor of the real magnetic component. This approach results in a simple first order transfer function, and therefore the tuning of the controller can be made very easily. After tuning the regulator, the duty ratio at the UM leg of the converter, D_{3_D}, can be obtained from Equation (29). After linearizing:

$$D_{3_D} = D_{1_D} + \frac{u_{UM} - u_{L_{UM}}}{V_{DC}} \tag{49}$$

Figure 13b shows the block diagram of the control scheme, where the measured DC link and UM voltage values, $V_{DC_{Meas}}$ and $U_{UM_{Meas}}$, respectively, are used to compute D_{3_D}. The implemented filter $H(s)$ is a second order Butterworth filter, on a Sallen–Key configuration, with a cut-off frequency of 3.5 kHz. The chosen bandwidth of the PI regulator $R(s)$ is BW = 300 Hz. The UM inductor has an inductance value of L_{UM} = 21 mH and a series parasitic resistor of R_{UM} = 0.48 Ω. Figure 14 shows the open loop gain of $G(s) \cdot R(s) \cdot H(s)$, used to check the system stability. As can be seen, for this design, the Phase Margin (PM) is close to 90°, therefore the system is stable.

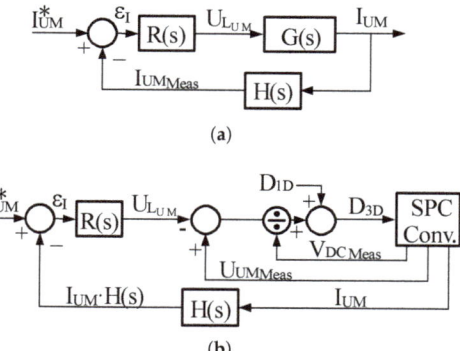

Figure 13. Control schemes: (**a**) Current control loop simplified scheme for tuning the regulator; and (**b**) implemented control scheme, obtaining D_{3_D} from the control action, $U_{L_{UM}}$.

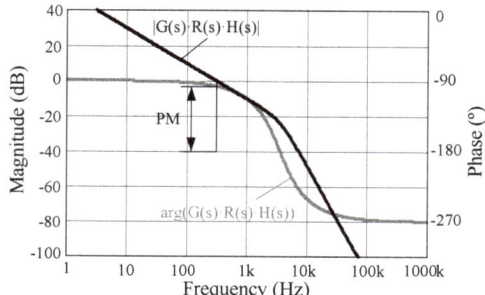

Figure 14. Bode plot with the Phase Margin (PM) of the system.

10. Conclusions and Future Developments

In this work, a comparison among basic power converter topologies for multiport Hybrid Storage Systems (HSSs) has been conducted. The conditions in which the results of the analysis are valid have been clearly defined. These particular constraints include the target application HSS, with a DC link interfaced with two storage units, one of them with significantly smaller voltage ratings. In addition, the constraints consider low/medium power levels where galvanic isolation is not a requirement. The control scheme implemented in the analysis manages the storage power flows to compensate the system DC link voltage due load steps or grid perturbations.

The topologies under study are the base Direct Parallel Connection (DPC) of two bidirectional converters, the Full-Bridge Converter (FBC) configuration for the UM leg, the Series Connection (SC) of the storage devices, and the proposed Series-Parallel Connection (SPC) of the storage units. All these options can be implemented by connecting the power switches in the standard single-leg configuration with complementary control pulses switching. This constraint facilitates the final implementation using ready-to-market, cheap components.

The results of the theoretical study, that included aspects such as losses, efficiency, loss balance between switches, and margins in the control stage design, have been validated by means of simulations and experimental tests on a built laboratory prototype with a rated power level of 10 kW. All throughout the analysis, and for the purpose of finding comparable results, the main parameters in the design, i.e., power, voltage and current levels of the devices were kept constant. For the same reason, also the values of the reactive elements were kept constant.

For HSS applications, SPC presents better efficiency (fewer losses) and also a better distribution of the electrical and thermal stresses in the switches of the legs of the converter. The combination of both effects yield to an increase in the reliability of the system.

The conclusions to this study are shown in Table 6 for DPC, FBC and SPC options. SC has been discarded, as it does not allow the use in HSS due the limitations in the controllability of the storage devices.

Table 6. Configuration performance.

Parameter	DPC	FBC	SPC
Efficiency	Baseline for comparison	Smaller than DPC	Depends on currents sign
Electr. and therm. stress balancing	High mismatch in K_f at switches	K_f at switches evenly distributed	K_f at switches evenly distributed
Control regulation margins	Non-symmetric current control, lim. bandwidth	Symmetrical current control	Symmetrical current control
Control simplicity	Simple, independent current control for EB and UM		
Current ripple through UM	Baseline for comparison at switching frequency	Ripple at twice the switching frequency	Ripple at twice the switching frequency
Current ratings at EB leg switches	Rated for EB peak current	Rated for EB peak current	Rated for algebraic sum at UM and EB peak currents
Size	Baseline for comparison	Increased No. of legs	Same legs than DPC, smaller UM inductor

From the comparison, it can be seen how SPC presents a better electric and thermal stresses balancing than the DPC case. Given that the UM inductor presents half the inductance value than in the DPC case for the same target current ripple, higher power density might also be achieved. SPC also allows for extended control margin. On the other hand, as mentioned, SC does not allow for an independent current control of the storage devices, therefore preventing its use as hybrid storage solution. For SC and FBC cases, the thermal and stresses balance is similar to the SPC, however FBC presents increased power losses vs. SPC.

The comparative efficiency results show how the performance comparison between DPC and SPC depend on the signs of the currents; therefore, the control scheme determines the overall efficiency of the system. From the above discussion, the proposed SPC scheme is considered as a feasible option for non-isolated interfacing of highly mismatched voltage rating storage systems in multiport configurations, for low to medium power rated HSSs applications.

The key issue demonstrated in this work is that, regardless of the efficiency of the base-case (DPC), an increase in the efficiency, in the dynamic performance, and in the stresses distribution in the converter switches achieved by the SPC connection scheme will be obtained, provided that a set of operating constraints are met. In the performed comparison, the hardware setup has been kept constant, and therefore this gain does not yield to modification in the components count or in the basic control implementation requirements.

Future developments include the optimization of the system for increasing the efficiency and power density assuming SPC scheme as the target topology; the inclusion of the energy storage devices modeling as to refine the control algorithms performance; or extension of the proposed solution in other kind of applications apart from HSS.

Acknowledgments: This work has been partially supported by the Spanish Government, Innovation Development and Research Office (MEC), under research grant ENE2016-77919, Project "Conciliator", and by the European Union through ERFD Structural Funds (FEDER). This work has been partially supported by the government of Principality of Asturias, Foundation for the Promotion in Asturias of Applied Scientific Research and Technology (FICYT), under Severo Ochoa research grants, PA-13-PF-BP13138 and PF-BP14135.

Author Contributions: Ramy Georgious and Jorge Garcia conceived the research and designed the experiments. Ramy Georgious performed the experiments and wrote the paper. All authors analyzed the data, and contributed in the discussion and conclusions

Conflicts of Interest: The authors declare no conflict of interest. The founding sponsors had no role in the design of the study; in the collection, analyses, or interpretation of data; in the writing of the manuscript, and in the decision to publish the results.

Abbreviations

The following abbreviations are used in this manuscript:

DPC	Direct Parallel Connection
EB	Electrochemical Battery
ESS	Energy Storage System
FBC	Full-Bridge Converter
K_f	Form Factor
HSS	Hybrid Storage Systems
PEC	Power Electronic Converter
PECG	Grid-tied Power Electronic Converter
PECL	Power Electronic Converter at Load/Generator
SC	Series Connection
SPC	Series-Parallel Connection
UM	Ultracapacitor Module

References

1. Thounthong, P.; Rael, S. The benefits of hybridization. *IEEE Ind. Electron. Mag.* **2009**, *3*, 25–37.
2. Jayasinghe, S.D.G.; Vilathgamuwa, D.M. Flying Supercapacitors as Power Smoothing Elements in Wind Generation. *IEEE Trans. Ind. Electron.* **2013**, *60*, 2909–2918.
3. Wu, H.; Xu, P.; Hu, H.; Zhou, Z.; Xing, Y. Multiport Converters Based on Integration of Full-Bridge and Bidirectional DC-DC Topologies for Renewable Generation Systems. *IEEE Trans. Ind. Electron.* **2014**, *61*, 856–869.
4. Abeywardana, D.B.W.; Hredzak, B.; Agelidis, V.G. Single-Phase Grid-Connected LiFePO/Battery-Supercapacitor Hybrid Energy Storage System With Interleaved Boost Inverter. *IEEE Trans. Power Electron.* **2015**, *30*, 5591–5604.

5. Pereirinha, P.G.; Trovao, J.P. Multiple Energy Sources Hybridization, The Future of Electric Vehicles? In *New Generation of Electric Vehicles*; Stevic, Z., Ed.; InTech: London, UK, 2012.
6. Lu, S.; Corzine, K.A.; Ferdowsi, M. A Unique Ultracapacitor Direct Integration Scheme in Multilevel Motor Drives for Large Vehicle Propulsion. *IEEE Trans. Veh. Technol.* **2007**, *56*, 1506–1515.
7. Abdel-baqi, O.; Nasiri, A.; Miller, P. Dynamic Performance Improvement and Peak Power Limiting Using Ultracapacitor Storage System for Hydraulic Mining Shovels. *IEEE Trans. Ind. Electron.* **2015**, *62*, 3173–3181.
8. Liu, J.; Zhang, L. Strategy Design of Hybrid Energy Storage System for Smoothing Wind Power Fluctuations. *Energies* **2016**, *9*, 991.
9. Miñambres Marcos, V.M.; Guerrero-Martínez, M.A.; Barrero-González, F.; Milanés-Montero, M.I. A Grid Connected Photovoltaic Inverter with Battery-Supercapacitor Hybrid Energy Storage. *Sensors* **2017**, *17*, 1856.
10. Jayasinghe, S.D.G.; Vilathgamuwa, D.M.; Madawala, U.K. A direct integration scheme for battery-supercapacitor hybrid energy storage systems with the use of grid side inverter. In Proceedings of the 2011 Twenty-Sixth Annual IEEE Applied Power Electronics Conference and Exposition (APEC), Fort Worth, TX, USA, 6–11 March 2011; pp. 1388–1393.
11. Guidi, G.; Undeland, T.M.; Hori, Y. An Optimized Converter for Battery-Supercapacitor Interface. In Proceedings of the 2007 IEEE Power Electronics Specialists Conference, Orlando, FL, USA, 17–21 June 2007; pp. 2976–2981.
12. Barrade, P.; Delalay, S.; Rufer, A. Direct Connection of Supercapacitors to Photovoltaic Panels With On-Off Maximum Power Point Tracking. *IEEE Trans. Sustain. Energy* **2012**, *3*, 283–294.
13. Yoo, H.; Sul, S.K.; Park, Y.; Jeong, J. System Integration and Power-Flow Management for a Series Hybrid Electric Vehicle Using Supercapacitors and Batteries. *IEEE Trans. Ind. Appl.* **2008**, *44*, 108–114.
14. Oriti, G.; Julian, A.L.; Anglani, N.; Hernandez, G.D. Novel Hybrid Energy Storage Control for a Single Phase Energy Management System in a Remote Islanded Microgrid. In Proceedings of the 2017 IEEE Energy Conversion Congress and Exposition (ECCE), Cincinnati, OH, USA, 1–5 October 2017; pp. 1552–1559.
15. Mohan, N.; Undeland, T.M.; Robbins, W.P. *Power Electronics: Converters, Applications, and Design*, 3rd ed.; Wiley and Sons: Hoboken, NJ, USA, 2003.
16. Yamamoto, K.; Imakiire, A.; Lin, R.; Iimori, K. Comparison of configurations of voltage boosters in PWM inverter with voltage boosters with regenerating circuit augmented by electric double-layer capacitor. In Proceedings of the 2009 International Conference on Electrical Machines and Systems, Tokyo, Japan, 15–18 November 2009; pp. 1–6.
17. Tucker, J. *Understanding Output Voltage Limitations of DC/DC Buck Converters*; Texas Instruments Incorporated: Dallas, TX, USA, 2008.
18. Song, J.; Kwasinski. A. Analysis of the effects of duty cycle constraints in multiple-input converters for photovoltaic applications. In Proceedings of the INTELEC 2009—31st International Telecommunications Energy Conference, Incheon, Korea, 18–22 October 2009; pp. 1–5.
19. Dusmez, S.; Hasanzadeh, A.; Khaligh, A. Comparative Analysis of Bidirectional Three-Level DC-DC Converter for Automotive Applications. *IEEE Trans. Ind. Electron.* **2015**, *62*, 3305–3315.
20. Grbovic, P.J.; Delarue, P.; Moigne, P.L.; Bartholomeus, P. A Bidirectional Three-Level DC-DC Converter for the Ultracapacitor Applications. *IEEE Trans. Ind. Electron.* **2010**, *57*, 3415–3430.
21. Azongha, S.; Liu, L.; Li, H. Utilizing ultra-capacitor energy storage in motor drives with cascaded multilevel inverters. In Proceedings of the 2008 34th Annual Conference of IEEE Industrial Electronics, Orlando, FL, USA, 10–13 November 2008; pp. 2253–2258.
22. Rufer, A.; Barrade, P. A supercapacitor-based energy-storage system for elevators with soft commutated interface. *IEEE Trans. Ind. Appl.* **2002**, *38*, 1151–1159.
23. Ding, Z.; Yang, C.; Zhang, Z.; Wang, C.; Xie, S. A Novel Soft-Switching Multiport Bidirectional DC-DC Converter for Hybrid Energy Storage System. *IEEE Trans. Power Electron.* **2014**, *29*, 1595–1609.
24. Zhou, H.; Duong, T.; Sing, S.T.; Khambadkone, A.M. Interleaved bi-directional Dual Active Bridge DC-DC converter for interfacing ultracapacitor in micro-grid application. In Proceedings of the 2010 IEEE International Symposium on Industrial Electronics, Bari, Italy, 4–7 July 2010; pp. 2229–2234.
25. Tao, H.; Kotsopoulos, A.; Duarte, J.L.; Hendrix, M.A.M. Family of multiport bidirectional DC-DC converters. In *IEE Proceedings—Electric Power Applications*; IET: Stevenage, UK, 2006; Volume 153, pp. 451–458.

26. Georgious, R.; Garcia, J.; Navarro, A.; Saeed, S.; Garcia, P. Series-Parallel Connection of Low-Voltage sources for integration of galvanically isolated Energy Storage Systems. In Proceedings of the 2016 IEEE Applied Power Electronics Conference and Exposition (APEC), Long Beach, CA, USA, 20–24 March 2016; pp. 3508–3513.
27. Garcia, J.; Georgious, R.; Garcia, P.; Navarro-Rodriguez, A. Non-isolated high-gain three-port converter for hybrid storage systems. In Proceedings of the 2016 IEEE Energy Conversion Congress and Exposition (ECCE), Milwaukee, WI, USA, 18–22 September 2016; pp. 1–8.
28. Tofoli, F.L.; de Castro Pereira, D.; de Paula, W.J.; de Sousa Oliveira Júnior, D. Survey on non-isolated high-voltage step-up dc-dc topologies based on the boost converter. *IET Power Electron.* **2015**, *8*, 2044–2057.
29. Zhang, N.; Sutanto, D.; Muttaqi, K.M.; Zhang, B.; Qiu, D. High-voltage-gain quadratic boost converter with voltage multiplier. *IET Power Electron.* **2015**, *8*, 2511–2519.
30. Hu, X.; Gong, C. A High Voltage Gain DC-DC Converter Integrating Coupled-Inductor and Diode-Capacitor Techniques. *IEEE Trans. Power Electron.* **2014**, *29*, 789–800.
31. Silveira, G.C.; Tofoli, F.L.; Bezerra, L.D.S.; Torrico-Bascopé, R.P. A Nonisolated DC-DC Boost Converter With High Voltage Gain and Balanced Output Voltage. *IEEE Trans. Ind. Electron.* **2014**, *61*, 6739–6746.
32. Langarica-Córdoba, D.; Diaz-Saldierna, L.H.; Leyva-Ramos, J. Fuel-cell energy processing using a quadratic boost converter for high conversion ratios. In Proceedings of the 2015 IEEE 6th International Symposium on Power Electronics for Distributed Generation Systems (PEDG), Aachen, Germany, 22–25 June 2015; pp. 1–7.
33. Georgiev, A.; Papanchev, T.; Nikolov, N. Reliability assessment of power semiconductor devices. In Proceedings of the 2016 19th International Symposium on Electrical Apparatus and Technologies (SIELA), Bourgas, Bulgaria, 29 May–1 June 2016; pp. 1–4.
34. Wang, C.S.; Li, W.; Wang, Y.F.; Han, F.Q.; Meng, Z.; Li, G.D. An Isolated Three-Port Bidirectional DC-DC Converter with Enlarged ZVS Region for HESS Applications in DC Microgrids. *Energies* **2017**, *10*, 446.
35. Tan, X.; Li, Q.; Wang, H. Advances and trends of energy storage technology in Microgrid. *Int. J. Electr. Power Energy Syst.* **2013**, *44*, 179–191.

© 2018 by the authors. Licensee MDPI, Basel, Switzerland. This article is an open access article distributed under the terms and conditions of the Creative Commons Attribution (CC BY) license (http://creativecommons.org/licenses/by/4.0/).

Article

Optimal Scheduling of Hybrid Energy Resources for a Smart Home

Muhammad Kashif Rafique [1], Zunaib Maqsood Haider [1], Khawaja Khalid Mehmood [1], Muhammad Saeed Uz Zaman [1], Muhammad Irfan [2] and Saad Ullah Khan [1] and Chul-Hwan Kim [1,*]

1. Department of Electrical and Computer Engineering, Sungkyunkwan University, Suwon 16419, Korea; kashif@skku.edu (M.K.R); zmhaider@skku.edu (Z.M.H.); khalidmzd@skku.edu (K.K.M.); saeed568@skku.edu (M.S.U.Z.); saadkhan@skku.edu (S.U.K.)
2. Department of Electrical Engineering, Khwaja Fareed University of Engineering and Information Technology (KFUEIT), Rahim Yar Khan 64200, Pakistan; muhammad.irfan@kfueit.edu.pk
* Correspondence: chkim@skku.edu; Tel.: +82-31-290-7124

Received: 29 September 2018; Accepted: 14 November 2018; Published: 18 November 2018

Abstract: The present environmental and economic conditions call for the increased use of hybrid energy resources and, concurrently, recent developments in combined heat and power (CHP) systems enable their use at a domestic level. In this work, the optimal scheduling of electric and gas energy resources is achieved for a smart home (SH) which is equipped with a fuel cell-based micro-CHP system. The SH energy system has thermal and electrical loops that contain an auxiliary boiler, a battery energy storage system, and an electrical vehicle besides other typical loads. The optimal operational cost of the SH is achieved using the real coded genetic algorithm (RCGA) under various scenarios of utility tariff and availability of hybrid energy resources. The results compare different scenarios and point-out the conditions for economic operation of micro-CHP and hybrid energy systems for an SH.

Keywords: battery energy storage system (BESS); electric vehicle (EV); fuel cell (FC); micro combined heat and power (micro-CHP) system; real coded genetic algorithm (RCGA); smart home (SH)

1. Introduction

1.1. Background and Motivation

The share of non-electric energy resources such as natural gas in modern power systems is significantly increasing due to environmental, economic and reliability concerns. At the same time, penetration of state of the art combined heat and power (CHP) systems is also on the rise due to their high efficiency and compact size. A CHP system is a cogeneration system that provides heat and electricity simultaneously. The recent technological advancement has made possible the miniaturization of cogeneration systems into micro-CHP units and their integration to power networks at a smart home (SH) level. A report from American Council for an Energy-Efficient Economy reveals that the CHP systems can operate at a high efficiency (80%) in comparison to the conventional modes of separately producing heat and power at a low efficiency (45%) [1]. Another major development in recent years is the increased penetration of low-emission electric vehicles (EV), which are significantly less dependent on the scarce fossil fuels. EVs, however, pose a challenge to the stability and economy of power systems as they require a plentiful power for their battery charging. Considering the high penetration of EVs and micro-CHP systems at house level, there is a need of comprehensive research to discover their optimal utilization and in-sync operation with the utility grid. Moreover, the feasibility of an integrated operation of the EVs and the CHP systems will increase if the economy of their combined operation is studied carefully [2]. So, this study presents an optimal scheduling of electrical and gas energy resources for a house in the presence of an EV.

1.2. Literature Review

CHP systems are an interesting topic among researchers, and a significant work has been devoted to study their feasibility, operation and to address the associated challenges [3–6]. A review of micro-CHP systems for residential applications concluded that 30% CO_2 emission can be reduced using micro-CHP systems [7]. In [8], a cost saving of 29% is achieved after applying a stochastic programming based reliability constrained optimization approach to CHP system components. A discrete optimization model for the optimal operation of a CHP system composed of a gas turbine and an auxiliary boiler is presented by Xie et al. in [9]. The study concluded that the increased CHP loading may not always result in an economic operation and the thermal to electric ratio also affects the profit of a CHP system. Different design options for integration of fuel cell (FC)-based micro-CHP systems in residential buildings is presented in [10]. Zhi et al reported that a gas turbine based combined cooling, heating and power system including electric batteries has multiple advantages, but the efficiency of the system decreases gradually with load reduction [11]. References [12,13] presented the economic operation of an FC-based CHP system in which different scenarios of recovered heat dissipation were compared.

Romano et al. [14] designed a Monte Carlo simulation based hybrid energy management system (EMS) having a PV and a battery energy storage system (BESS) in a smart house. The EMS was used to control the schedulable loads. A dynamic simulation was performed to study the interaction between and internal combustion engine (ICE) based micro-CHP system and the EV charging in a semi-detached home in two different geographical locations in Italy [15]. A parametric analysis based on different daily driving distances for EVs was performed and the proposed method resulted a cost saving of up to 60%.

The combined use of a plug-in hybrid electric vehicle (PHEV) and a polymer electrolyte fuel cell-based cogeneration system is discussed in [16], and the developed model was analyzed using mixed-integer linear programming. Due to an increased electric capacity factor and a thermal power supply rate, this synergistic operation resulted in energy saving and cost reduction as compared to their separate use. The combined impact of an FC-based micro-CHP system and PHEV on the annual utility energy consumption is studied in [17] for three daily running distances of 0, 10 and 20 km. The results show that the combined use of micro-CHP and PHEV reduced the annual utility consumption up to 3.7% compared to their separate use.

The potential synergy of the ICE-based micro-CHP and EV charging is explored in [18], and the results indicate improvement in the economy. The work in [19] presents the optimal charging and discharging schedule of EVs in a parking station installed with the PV and BESS to minimize the overall operational cost.

A closed-form solution is proposed to schedule responsive loads with a special focus on the EV charging with uncertain departure times in [20]. In [21], an intelligent charging method for EVs is proposed considering time-of-use (TOU) tariff and in [22], considering the intermittent renewable energy sources. The work presented in [23] proposed a rule-based energy management scheme considering flat rates of the electricity.

Wu et al. proposed a scheme for the cost minimization of electricity considering the power demands of a home and EV charging [24]. However, the thermal loads of the home were not taken into account in the study. García-Villalobos et al. [25] reviewed different PHEV charging strategies (i.e., charging without any special control, charging during off-peak period, valley filling charging, and peak shaving charging). The findings of this study suggest that although the former two techniques are user-friendly and easy to implement, the latter two methods result in improved ancillary services, flattened load profile, and optimal integration of renewable energy resources. An EMS to regulate voltage profile and allocate power shares to EVs is proposed in [26]. The residential EV charging impact upon the distribution system voltages was reviewed in [27], and a method was proposed to mitigate the EV load effects. Infrastructural changes as well as TOU-pricing based indirect EV charging controls were proposed, and an optimal TOU schedule was presented with the objective of maximizing both the utility and the customer benefits. In [28], an optimization scheme is proposed to schedule the household appliances in an SH network. The above referenced studies provide a valuable contribution to the literature, however, in the context of modern SHs which are equipped with hybrid energy resources and EVs, following aspects need more attention.

- Although a significant research has been performed that deals independently with micro-CHP systems and EVs (e.g., [7,12,13,24]), their combined operation needs more attention as technological advancements envision their integrated utilization at homes.
- A comprehensive economic analysis and scheduling of electrical and thermal loads are not provided (e.g., [10,16,17,29]). The feasibility of EV integration into a micro-CHP system will be increased if its economic operation is analyzed carefully.
- The inclusion of an EV exerts a unique stress on the house loads. It raises the electrical demand while the thermal load remains unchanged. Hence, the feasibility of its responsive behavior must be explored.

1.3. Contribution and Paper Organization

The contribution and highlights of this work are summarized as follows:

1. A model of an SH is developed. The SH is equipped with an EV, a BESS, and an FC-based micro-CHP system which is powered by natural gas. Two typical tariffs (flat and variable) of the utility are realized, and the effect of responsive nature of the EV is explored.
2. An optimization problem for the economic operation of the hybrid energy system of the SH is developed, and the constraints are defined for the systems and the devices. The problem is modeled to utilize the real coded genetic algorithm (RCGA) to optimally schedule the resources and the responsive loads.
3. A comprehensive simulation results of six test cases reveal interesting features of the developed model and optimization process. The necessary conditions for the optimal operation of the energy resources are also discussed.

The remaining paper is organized as follows. An SH model is developed in Section 2, and an optimization problem and constraints are defined in Section 3. Section 4 explains the RCGA, and Section 5 discusses the test cases and simulation results. Finally, the conclusion is presented in Section 6.

2. Development of SH Model

A typical SH having hybrid energy resources such as natural gas and electric power is presented in Figure 1. The energy flow in the SH is divided into two loops.

Thermal Loop:

It consists of thermal loads, an auxiliary boiler, and recovered heat from an FC. The energy source for the FC and auxiliary boiler is natural gas. The waste heat from the FC is recovered and provided to the thermal loads. If the heat provided by the FC is not sufficient to meet the total thermal power demand, the deficit power is provided by the auxiliary boiler.

Electrical Loop:

It consists of responsive and non-responsive electrical loads and electrical power resources (i.e., utility, FC, and BESS). The charging and discharging of the BESS depends upon its efficiency and the energy already stored in it according to the developed optimization model (as explained in Section 4). Modeling of the components of an SH is presented as follows.

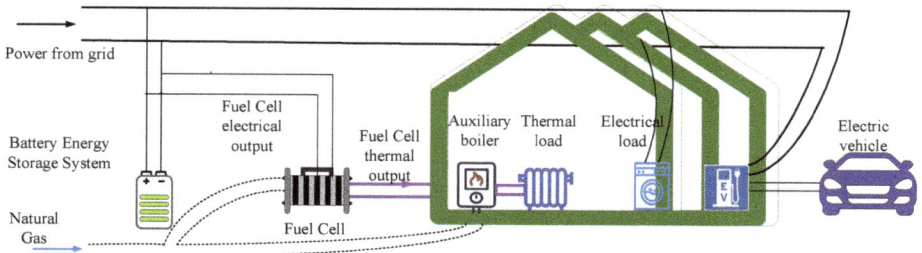

Figure 1. Overview of a smart house.

2.1. FC Model

There are several types of FCs depending upon the fuel used for energy conversion. In this work, a proton exchange membrane fuel cell (PEM-FC) [30,31] is used. Serving as a micro-CHP system for the SH, input of the FC is natural gas and its outputs are electricity and heat. Efficiency of the FC is related to its part load ratio (PLR). PLR is the ratio of the electrical output at interval i to maximum power rating of the FC and is given in Equation (1):

$$PLR_i = P_{FC_e,i}/P_{FC_{max}} \quad (1)$$

where PLR_i is the PLR at interval i when the FC output power is $P_{FC_e,i}$. Mathematical relations for the efficiency and thermal to electric ratio (rTE) are given as follows.

When $PLR_i < 0.05$:

$$\eta_{FC,i} = 0.2716; \quad r_{FC,i} = 0.6816 \quad (2)$$

When $PLR_i > 0.05$:

$$\eta_{FC,i} = 0.9033 PLR_i^5 - 2.9996 PLR_i^4 + 3.6503 PLR_i^3 - 2.0704 PLR_i^2 + 0.4623 PLR_i + 0.3747 \quad (3)$$

$$r_{TE,i} = 1.0785 PLR_i^4 - 1.9739 PLR_i^3 + 1.5005 PLR_i^2 - 0.2817 PLR_i + 0.6838 \quad (4)$$

Having $r_{TE,i}$, the thermal power ($P_{FC_h,i}$) produced by the FC at interval i is calculated as:

$$P_{FC_h,i} = r_{TE,i} P_{FC_e,i} \quad (5)$$

Figure 2 represents typical performance characteristics of an FC [32]. At very low PLR (<10%), the parasitic losses are high and the overall efficiency is very low. Beyond this region, the FC operates at 30–40% electrical efficiency. The efficiency is slightly higher at lower PLR compared to the peak power operation. However, the performance and efficiency vary for different designs of FCs.

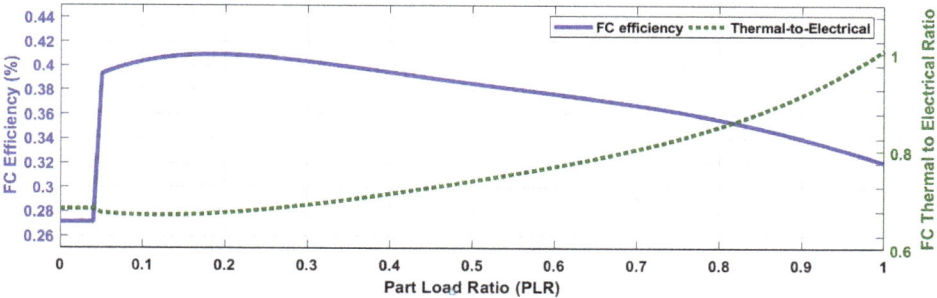

Figure 2. Fuel cell efficiency and thermal to electrical ratio as function of Part Load Ratio.

2.2. EV Model

To model the battery of an EV, several factors are required to be considered. These factors include the state of charge (SOC) at plug-out time, driving distance, driving style, route choice, traffic, etc. In this study, the data available in [33,34] is used. The effect of driving distance on SOC of battery is modeled as follows [24]:

$$SOC_{EV.pi} = \begin{cases} SOC_{EV_{min}} & \text{if } \left(SOC_{EV.po} - \frac{d}{\eta_{EV} EV_{cap}}\right) \leq SOC_{EV_{min}} \\ SOC_{EV.po} - \frac{d}{\eta_{EV} C_{EV}} & \text{Otherwise} \end{cases} \quad (6)$$

where

P_{EV_i}	EV charging power at interval i (kW)
$SOC_{EV.i}$	EV SOC at interval i (%)
$SOC_{EV_{pi}}$	EV SOC at plugging-in time (%)
$SOC_{EV_{po}}$	EV SOC at plugging-out time (%)
$SOC_{EV_{min}}$	Minimum SOC of the EV (%)
d	Trip distance of the EV ($miles$)
η_{EV}	Overall electric drive efficiency
C_{EV}	Capacity of the EV (kWh)

If d and $SOC_{EV_{po}}$ are given, the $SOC_{EV_{pi}}$ can be calculated using Equation (6). The $SOC_{EV_{pi}}$ is lower-bounded by the minimum SOC of the EV, which prevents the depletion damage to the EV battery.

The charging of EV is expressed as follows

$$SOC_{EV.i} = SOC_{EV.i-1} + \frac{P_{EV.i} \times T}{EV_{cap}} \times 100 \quad (7)$$

2.3. BESS Model

The model of a BESS is provided in Equation (8). The charging and discharging efficiencies of the battery are considered in this work and the net BESS efficiency is 90%.

$$W_{B.i} = W_{B.i-1} + \left[T\eta_{B.ch} \quad -\frac{T}{\eta_{B.dch}}\right]\mu_i \quad (8)$$

where $W_{B.i}$ is energy of the BESS at interval i, and $\mu_i = \begin{bmatrix} P_{B.i,ch} \\ P_{B.i,dch} \end{bmatrix}$ denotes the power vector with the BESS charging and discharging powers, $\eta_{B.ch}$ and $\eta_{B.dch}$ are the charging and discharging efficiencies of the BESS, and T is the simulation step.

2.4. TOU Tariff

A TOU tariff refers to the different prices of electricity at different hours. Typically, the power demand is higher during certain time intervals of a day causing the overloading of a power grid. The utility companies set a higher price of electricity during these intervals to reduce stress on the power system. On the other hand, a lower price at some other intervals attracts the consumers and improves the utilization factor. In this work, the tariff considered for the SH is a peak-valley tariff which is a type of the TOU tariff. Three different prices of electricity are considered in this work during peak, plain and valley hours [35]. The price of electricity during these intervals is listed in Table 2 with their corresponding time intervals [36]. The price is normalized with respect to the maximum price defined in the peak period. These normalized values are used in Equation (12).

Table 1. Peak-Valley Electricity Tariff.

Tarrif Type	Time Range	Normalized Price $T_{p.v}$ ($p.u$)
Peak	[09:00–12:00] [17:00–22:00]	1
Plain	[13:00–16:00]	0.9
Valley	[01:00–08:00] [23:00–24:00]	0.78

3. Optimization Model

This section presents the optimization model for the SH. The goal of the proposed optimization model is to minimize the 24 h operating costs of the SH subject to the following assumptions:

- The forecasted data for the thermal and electrical loads is available.
- The initial conditions of the *SOC* of the BESS and trip distance of the EV is available.
- All the devices are already installed. Therefore, the installation costs are not considered.

3.1. Objective Function

The objective function to be minimized is modeled as

$$\min \left[\sum_{i=1}^{n} (C_{FC,i} + C_{BL,i} + C_{U,i}) \right] \qquad (9)$$

where

$$C_{FC.i} = \begin{cases} T \cdot C_{gas} \left(\dfrac{P_{FC_e.i}}{\eta_{FC}.i} \right) + \alpha & \text{if } P_{FC_e.i-1} = 0 \\ & \text{and } P_{FC_e.i} > 0 \\ \beta & \text{if } P_{FC_e.i-1} > 0 \\ & \text{and } P_{FC_e.i} = 0 \\ T \cdot C_{gas} \left(\dfrac{P_{FC_e.i}}{\eta_{FC}.i} \right) & \text{else} \end{cases} \qquad (10)$$

$$C_{BL.i} = T \cdot C_{gas} \cdot P_{BL.i} \qquad (11)$$

$$C_{U.i} = T \cdot T_{p.v} \cdot C_{U_b} \cdot P_{U.i} \qquad (12)$$

where

n	Number of hours
T	Length of a time interval (h)
α, β	Startup, Shutdown costs of the FC
$C_{FC.i}$	Cost of the FC operation for interval i ($/kWh)
$C_{BL.i}$	Cost of the boiler operation for interval i ($/kWh)
$C_{U.i}$	Cost of the utility power for interval i ($/kWh)
C_{gas}	Cost for purchasing the gas ($/kWh)
C_{U_b}	Base cost for purchasing the power from utility
$P_{FC_e.i}$	Electrical power from the FC at interval i (kW)
$P_{BL.i}$	Heating provided by the boiler at interval i (kW)
$P_{U.i}$	Electrical power provided by the utility at interval i (kW)
$T_{p.v}$	Multiplier for the peak-valley price as provided in Table 1
$\eta_{FC.i}$	Efficiency of the FC

3.2. Constraints

Due to physical and operational limits of the devices and energy systems, the variables for power, energy and SOC should meet the following constraints during the optimization process.

3.2.1. Constraints of Power Balance

Electrical Power Balance

The input power from the utility is distributed among the electrical loads. The BESS either works as a source of electric power or an electric load. Therefore, following equations model the electrical power balance and this dual role of BESS.

When the BESS is charging:

$$P_{D_e.i} + P_{EV.i} - P_{FC_e.i} + \frac{P_{B.i}}{\eta_{ch}} - P_{U.i} = 0 \tag{13}$$

When the BESS is discharging:

$$P_{D_e.i} + P_{EV.i} - P_{FC_e.i} + \eta_{dch} P_{B.i} - P_{U.i} = 0 \tag{14}$$

where

$P_{D_e.i}$ Electrical demand at interval i (kW)
$P_{EV.i}$ Power being delivered to the EV at interval i (kW)
$P_{B.i}$ BESS power at interval i (kW). It is negative in charging mode and positive in discharging mode
$P_{D_h.i}$ Heating demand at interval i (kW)
$P_{FC_h.i}$ Heating produces by the FC at interval i (kW)
η_{ch} Charging efficiency of the BESS ($p.u$)
η_{dch} Discharging efficiency of the BESS ($p.u$)

Thermal Power Balance

The total demand of thermal power is met by the FC and the auxiliary boiler in the SH. The constraint of thermal power balance is formulated as:

$$P_{D_h.i} - P_{FC_h.i} - P_{BL.i} = 0 \tag{15}$$

where

$P_{D_h.i}$ Heating demand at interval i (kW)
$P_{FC_h.i}$ Heating produced by the FC at interval i (kW)
$P_{BL.i}$ Heating produced by the auxiliary boiler at interval i (kW)

3.2.2. Constraints of Devices

The constraints applicable to the devices available in the SH are explained below.

Constraints of FC

The rate of change of the FC output is limited to the upper and the lower boundaries of ramp rate. Therefore, following inequalities must be satisfied:

$$P_{FC_e.i} - P_{FC_e.i-1} < \Delta P_{FC_{up}} \tag{16}$$

$$P_{FC_e.i-1} - P_{FC_e.i} < \Delta P_{FC_{dn}} \tag{17}$$

$$P_{FC_{min}} < P_{FC_e.i} < P_{FC_{max}} \tag{18}$$

where

$\Delta P_{FC_{up}}$	FC ramp rate limit for increasing power
$\Delta P_{FC_{dn}}$	FC ramp rate limit for decreasing power
$P_{FC_{min}}$	FC minimum power limit
$P_{FC_{max}}$	FC maximum power limit

Constraints of EV

Charging and discharging of the EV battery is subject to certain limitations regarding its maximum charging power and the SOC as given below:

$$P_{EV.i} < P_{EV_{chmax}} \tag{19}$$

$$SOC_{EV_{min}} \leq SOC_{EV.i} \leq SOC_{EV_{max}} \tag{20}$$

where $P_{EV_{chmax}}$ is the maximum charging power of the EV in (kW) and $SOC_{EV_{max}}$ is the maximum SOC.

Constraints of BESS

Following constraints the BESS must be satisfied:

$$W_{B_{min}} < W_{B.i} < W_{B_{max}} \tag{21}$$

If the battery is in charging/discharging mode, it is subjected to the maximum charging and discharging rates as explained below:

During Charging Mode

$$W_{B.i} - W_{B.i-1} < P_{B_{chmax}} \times T \tag{22}$$

During Discharging Mode

$$W_{B.i-1} - W_{B.i} < P_{B_{dchmax}} \times T \tag{23}$$

where

$W_{B.i}$	BESS energy at interval i (kWh)
$W_{B_{min}}$	BESS minimum energy limit (kWh)
$W_{B_{max}}$	BESS maximum energy limit (kWh)
$P_{B_{chmax}}$	BESS minimum charging rate limit (kW)
$P_{B_{dchmax}}$	BESS maximum discharging rate limit (kW)
T	Length of time interval

4. Real Coded Genetic Algorithm

Modern heuristic techniques are fast and emerging tools to optimize non-linear systems. Generally, they outperform the traditional derivative based techniques which have limitations of getting trapped in a local minimum, computational complexity, or are not applicable to certain objective functions. The Genetic Algorithm (GA) is one of the most used evolutionary algorithms in power system applications. Its mechanism is based on evolution in nature and the algorithm essentially consists of genetic operations of selection, cross-over and mutation applied to a population of chromosomes. RCGA which is an improved version of the GA is implemented in this study for the optimization purpose. For real valued numerical optimization problems, the floating point or integer representation of population variables in RCGA outperforms the binary representation of the variables in the GA. In comparison to the GA, the RCGA provides higher consistency, more precision and faster convergence [37].

RCGA is an efficient method which does not require a derivative of the objective function to find the optimal solution. Therefore, in contrast to the linear programming or derivative-based techniques, RCGA can effectively handle all types of objective functions and constraints whether they are smooth, non-smooth; linear, non-linear; continuous, discontinuous; convex, non-convex; stochastic or does not possess derivatives. A detailed discussion of the RCGA is available in [38–40]. A brief description of the steps involved in the implementation of the RCGA for optimal scheduling of the SH's energy resources is presented below.

4.1. Initialization

Similar to other evolutionary algorithms, the RCGA starts with generation of an initial population called "chromosomes". In an N-dimensional optimization problem, the position of i-th gene is determined as follows:

$$Chromosome_i = [x_1, x_2, x_3, \ldots, x_i, \ldots, x_N] \quad (i = 1, 2, \ldots, N_G)$$

where each gene denotes power (kW) of a device and N_G is total number of genes in a chromosome.

4.2. Dimensionality

The number of independent variables in a system determines the dimensions of an optimization problem. For the SH presented in this work, there are three independent variables namely P_{FC_e}, P_{EV} and P_{BT}. The information of these independent variables and power demands of the thermal and electrical loads are used to determine all the remaining unknown variables. For example, knowing P_{FC_e}, P_{FC_h} can be solved using Equations (2)–(5). Similarly, P_{BL} and P_U can be computed using Equations (13)–(15). The objective of this study is to calculate an optimal scheduling of SH devices for one day (24 h) with a time interval of $T = 1$ h. Therefore, three variables in each hour result in the dimension of optimization problem $N_G = 24 \times 3 = 72$.

If M is the number of chromosomes in one generation then $M \times 72$ gives the dimensionality in terms of one generation of the RCGA as shown in Figure 3.

$$\begin{matrix}
P_{EV_1}^1 & P_{EV_1}^2 & \ldots & P_{EV_1}^{24} & P_{FC_e1}^1 & P_{FC_e1}^2 & \ldots & P_{FC_e1}^{24} & P_{B_1}^1 & P_{B_1}^2 & \ldots & P_{B_1}^{24} \\
P_{EV_2}^1 & P_{EV_2}^2 & \ldots & P_{EV_2}^{24} & P_{FC_e2}^1 & P_{FC_e2}^2 & \ldots & P_{FC_e2}^{24} & P_{B_2}^1 & P_{B_2}^2 & \ldots & P_{B_2}^{24} \\
\vdots & \vdots & \vdots & \vdots & \vdots & \vdots & \vdots & \vdots & \vdots & \vdots & \vdots & \vdots \\
P_{EV_M}^1 & P_{EV_M}^2 & \ldots & P_{EV_M}^{24} & P_{FC_eM}^1 & P_{FC_eM}^2 & \ldots & P_{FC_eM}^{24} & P_{B_M}^1 & P_{B_M}^2 & \ldots & P_{B_M}^{24}
\end{matrix}$$

Figure 3. Chromosomes in one generation of the real coded genetic algorithm.

4.3. Implementation of the Constraints

In each time interval, all the system constraints for $P_{FC_e.i}$, $P_{BL.i}$ and $P_{EV.i}$ are checked as follows:

1. Constraints of FC

    - According to (16), if $P_{FC_e.i} > P_{FC_e.i-1}$ and $P_{FC_e.i} - P_{FC_e.i-1} > \Delta P_{FC_{up}}$, then $P_{FC_e.i}$ is assumed to be equal to $P_{FC_e.i-1} + \Delta P_{FC_{up}}$.
    - According to (17), if $P_{FC_e.i} < P_{FC_e.i-1}$ and $P_{FC_e.i-1} - P_{FC_e.i} > \Delta P_{FC_{dn}}$, then $P_{FC_e.i}$ is assumed to be equal to $P_{FC_e.i-1} - \Delta P_{FC_{dn}}$.
    - According to (18), in the case of $P_{FC_e.i} > P_{FC_{max}}$ or $P_{FC_e.i} < P_{FC_{min}}$, it is considered equal to $P_{FC_{max}}$ and $P_{FC_{min}}$ respectively.

2. Constraints of the BESS

    - For each interval i, if $W_{B.i}$ exceeds the battery capacity limit $W_{B_{max}}$ in charging mode i.e., $W_{B.ini} - \sum_{j=1}^{i} P_{B.j} > W_{B_{max}}$, then $P_{B.i} = W_{B_{max}} - W_{i-1}$ and $W_{B.i} = W_{i-1} - P_{B.i} \times 1$ h to satisfy the upper limit of (21).
    - For each interval i, if $W_{B.i}$ depletes more than battery minimum limit $W_{B_{min}}$ in discharging mode, i.e., $W_{B.ini} - \sum_{j=1}^{i} P_{B.j} < W_{B_{min}}$ then $P_{B.i} = W_{i-1} - W_{B_{min}}$ and $W_{B.i} = W_{i-1} - P_{B.i}$ to satisfy the lower limit of (21).
    - If the battery is in charging mode i.e., $P_{B.i} < 0$, then the difference between values of battery energy in two consecutive intervals $W_{B.i} - W_{B.i-1}$ should not exceed $P_{B_{chmax}}$ according to (22). Otherwise $W_{B.i} = W_{B.i-1} + P_{B_{chmax}} \times 1$ h.
    - If the battery is in discharging mode i.e., $P_{B.i} > 0$, then the difference between the values of battery energy in two consecutive intervals $W_{B.i-1} - W_{B.i}$ should be less than $P_{B_{dchmax}}$ according to (23). Otherwise $W_{B.i} = W_{B.i-1} + P_{B_{chmax}} \times 1$ h.

3. Constraints of the EV

    - To handle the EV constraints provided in Equations (19) and (20), a process similar to the one described for BESS is adopted.

The flowchart of the RCGA-based optimal scheduling process is shown in Figure 4.

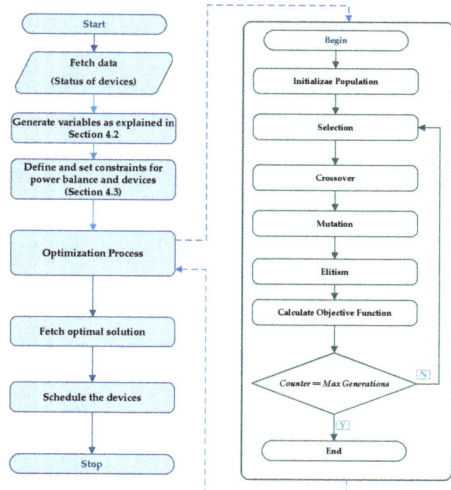

Figure 4. Flowchart of the proposed work.

5. Simulation Results

This section presents the results of the numerical simulations to highlight the significant features of the proposed optimization model. As the SH has hybrid energy resources and is equipped with various devices, therefore, multiple simulation scenarios are generated to compare their impacts under different utility tariffs as shown in Table 1. In all the simulation scenarios, the electric power can be purchased from the utility and the auxiliary boiler is available for the thermal energy. Table 2 presents these cases.

Table 2. Description of the test cases.

Case No	FC	EV	Variable Tariff	Scheduling of EV	BESS
1	x	x	x	x	x
2	o	x	x	x	x
3	o	o	x	x	x
4	o	o	o	x	x
5	o	o	o	o	x
6	o	o	o	o	o

The normalized curves for the 24 h thermal and electric power demands of the SH are shown in Figure 5 [41]. The thermal demand curve is relatively stable with mean to peak ratio of 91.5% whereas the electrical power demand curve is fluctuating and its mean to peak ratio is 83%. To meet the load demands, the proposed optimization model finds the optimal resources for the 24 h operation of the SH. The electrical resources need more attention due to significantly changing profile of the electrical power demand. The heat and electricity demands are 2.5 kW each for the SH. The cost is calculated for the 24 h. The micro-CHP system follows the electrical demand to generate the electricity while delivering the heat as a by-product. The EV used in this study is Mitsubishi's compact i-MiEV [42]. It can travel 62 miles on a full charge in typical driving conditions. The distance traveled by the EV, time-in and time-out are selected according to the U.S. National House-hold Travel Survey (NHTS) [33,34]. A smart charging mechanism for the EV and BESS is considered which charges them according to the optimized values generated by the proposed optimization model based on the RCGA. All the parameters related to the SH, EV, FC, BESS, and optimization model are given in Table 3.

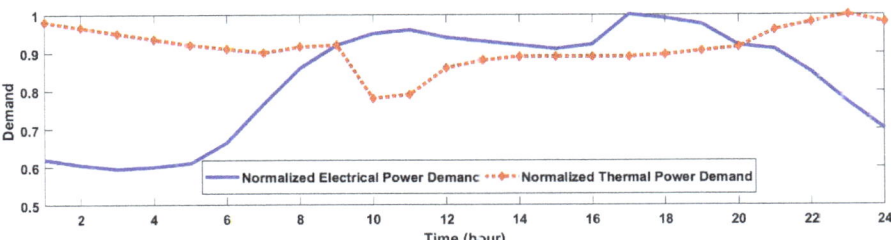

Figure 5. Daily thermal and electric power demands.

Table 3. System Parameters.

Parameter Description	Symbol	Value	Unit
Electric Vehicle			
Trip distance of EV	d	60	mi
Overall electric drive efficiecy	η_{EV}	6.2	-
Capacity of EV	C_{EV}	16	kWh
EV maximum charging power	$P_{EV_{chmax}}$	3.3	kW
Minimum SOC of EV	$SOC_{EV_{min}}$	3.3	%
Maximum SOC of EV	$SOC_{EV_{max}}$	100	%
EV SOC at plugging-out time	$SOC_{EV_{po}}$	100	%
Plug-in time	T_I	17:00	hour
Plug-out time	T_O	7:00	hour
Fuel Cell			
FC maximum power limit	$P_{FC_{max}}$	2	kW
FC minimum power limit	$P_{FC_{min}}$	0.05	kW
FC ramp rate limit for increasing power	$\Delta P_{FC_{up}}$	1.25	kW
FC ramp rate limit for decreasing power	$\Delta P_{FC_{dn}}$	1.5	kW
FC startup cost	α	0.15	$
FC shutdown cost	β	0	$
Battery Energy Storage System			
Maximum energy limit	$W_{B_{max}}$	3	kWh
Minimum energy limit	$W_{B_{min}}$	0	kWh
Minimum charging rate limit	$P_{B_{chmax}}$	c/4	kW
Maximum discharging rate limit	$P_{B_{dchmax}}$	c/2	kW
Charging efficiency of Battery	$\eta_{B.ch}$	0.927	-
Discharging efficiency of Battery	$\eta_{B.dch}$	0.971	-
General			
Number of hours	n	24	hour
Length of time interval	T	1	hour
Cost for purchasing gas	C_{gas}	0.05	$/kW
Base cost for purchsing power from utility	C_{U_b}	0.13	$/kW
Genetic Algorithm			
Crossover probability	Pc	0.5	-
Mutation probability	Pm	0.1	-

5.1. Case 1: Base Case

Case 1 serves as a base case which represents a typical conventional home without a micro-CHP and an EV. As shown in Table 3, this case does not consider availability of the BESS at the home, and the variable tariff is not applied. The proposed optimization model is not applicable to this case. The non-schedulable electric demand of the home is met by the utility, and the thermal demand is met by the boiler as shown in Figure 6. In this case, the total cost of the energy is 9.20 ($/day).

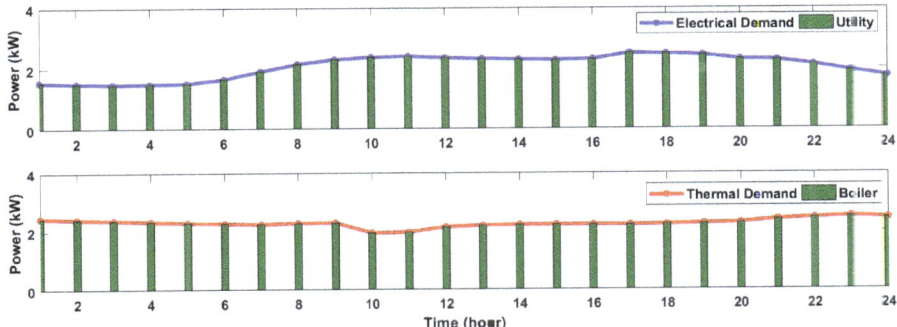

Figure 6. Case 1: Basic operation mode of house.

5.2. Case 2: Installation of FC

In this case, an FC is added to the SH which serves as a micro-CHP system, and a significant portion of electric and thermal loads shifts to it due to its economic operation. Due to the power rating of the FC, it cannot meet the whole electric power demand of the SH and a part of electric energy is purchased from the utility. Similarly, the auxiliary boiler provides the additional heating if the thermal power demand is more than the FC's thermal power output. Figure 7 shows the results for the optimal dispatch of the FC, utility and auxiliary boiler. It is observed that FC follows the electrical load curve in the night hours when demand is low. During the day, the FC works at near its maximum power generation limit. The remaining power is provided by the utility. No variable tariff is considered at this point, and the FC output is almost constant for 24 h. Another important point is that the FC is not working exactly at its maximum power generation capacity. According to (3), if the FC generates maximum power, its efficiency decreases and the total operational cost becomes higher than the cost of purchasing from the utility. The net cost is 7.97 ($/day) in this case which is 13.37% less as compared to the cost incurred in Case 1 when there was no micro-CHP in the SH.

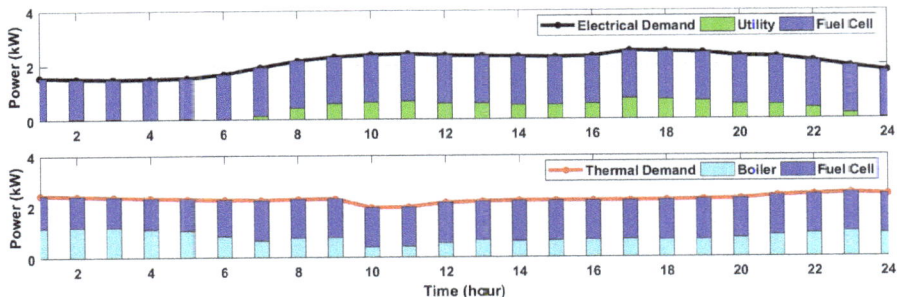

Figure 7 Case 2: Electrical and thermal demands after adding the fuel cell.

5.3. Case 3: Addition of EV

In this case, an EV is added to the SH and its impact is analyzed. This case considers the EV as a constant and unscheduled load. Thus, depending upon the initial SOC of EV, it presents itself as a constant load from the time it is plugged-in (T_I) until it gets fully charged. Figure 8 shows the SOC of EV battery and electric powers from the utility and FC in this case. The operating cost of the system is 9.98 ($/day) which is higher than Case 2 due to the loading effect of the EV.

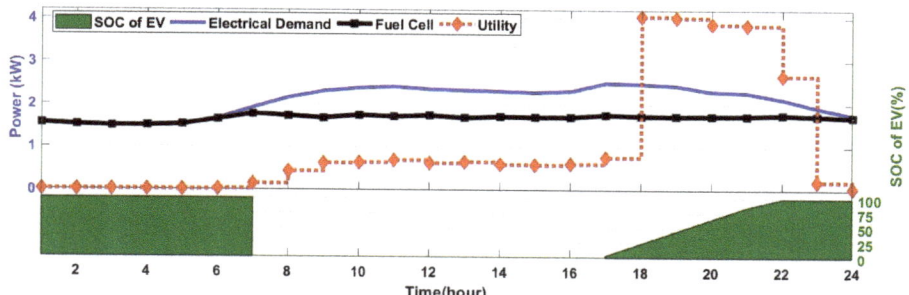

Figure 8. Case 3: Charging of electric vehicle (EV) without scheduling.

5.4. Case 4: Considering Variable Tariff

In the previous cases, the utility tariff was considered as flat rates for 24 h. This case and the following cases, however, consider a variable tariff which is widely applicable in the present power markets. A peak-valley tariff is considered in this case according to Table 1. As shown in Figure 9, EV loading on the system is the same as in Case 3 but the FC adjusts its output to take benefit of the valley prices. To achieve an economic operation, the proposed optimization algorithm makes the FC increase its output during peak price hours and reduce its output during valley price hours. This is in contrast to Case 2 when the FC was operating on almost constant output for 24 h. Due to the exploitation of variable tariff by the optimization model, the total system cost has decreased to 9.88 ($/day).

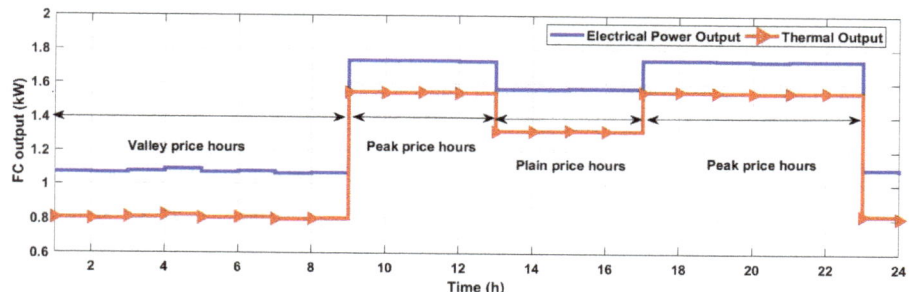

Figure 9. Case 4: Impact of peak-valley tariff on fuel cell output.

5.5. Case 5: Scheduling the EV Charging

The modern concept of scheduling of responsive electric loads results in great economic and technical benefits [43,44]. The electric loads of high power rating and low criticality such as EV, washing machines are the prime candidates for such scheduling. In this case, the EV is modeled as a responsive load and its charging is scheduled. The optimization algorithm selects those hours for the EV charging when the total cost is optimized and the EV is fully charged. Figure 10 shows the EV charging hours, *SOC* of the EV battery, and electric powers from the utility and the FC in this case.

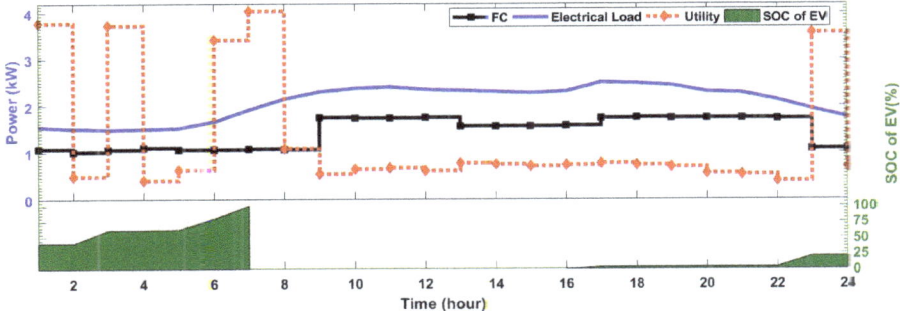

Figure 10. Case 5: Results of the system with scheduling of EV.

Although EV is connected to the system as soon as it reaches the house at 5:00 P.M., the optimization algorithm forced it to charge during those hours when the customer can get more benefit. The charging of EV starts from 11:00 P.M. (during valley prices) and the SOC reaches up to 100% before EV leaves home at 7:00 A.M. The cost reduces to 9.44 ($/day) which is around 5% decrease as compared to Case 4.

5.6. Case 6: Adding the BESS

This case considers the addition of a BESS to the SH and analyzes its impacts. It is notable that using the BESS with the utility can result in an economic operation only if the product of charging and discharging efficiencies is greater than the valley-to-peak price ratio. In this case, BESS is charged during the valley price hours and discharged during the peak price hours. Therefore, it is pertinent to introduce BESS efficiency in this section which is $\eta_B = \eta_{B.ch} \times \eta_{B.dch}$. In this work, BESS efficiency (η_B) is 0.9, and charging and discharging of BESS is shown in Figure 11. The product of the two efficiencies is more than the valley-to-peak price ratio of the utility tariff therefore energy routing through BESS can result in economic operation. However, to exploit the maximum benefit from BESS, the proposed optimization algorithm selects its charging and discharging hours as explained below.

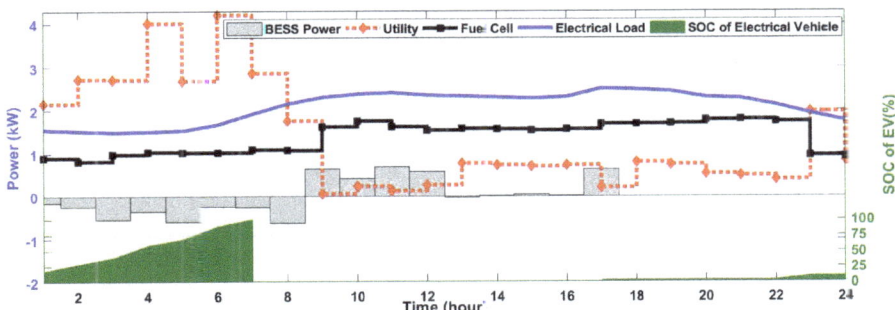

Figure 11. Case 6: Results of the simulation after adding battery energy storage system.

The utility tariff is at its lowest price during 01:00–08:00, therefore BESS is charged. The FC does not operate at its maximum because it generates electricity such that its cost per unit is lower than or equal to utility. Energy share provided by the utility is maximum during this interval. The charging of BESS at this tariff results in 0.8667 p.u/kW of the dischargeable energy. For thermal demand, most of the heating is provided by the boiler, and FC adds to some extent as shown in Figure 12.

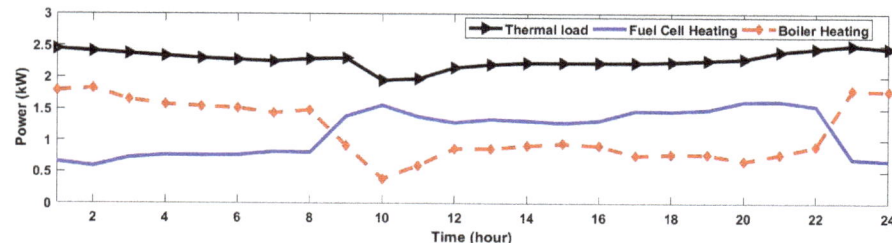

Figure 12. Case 6: Thermal heating.

From 09:00–12:00, utility tariff is at its maximum price. The proposed optimization algorithm results in low purchase from the utility, discharging of the BESS to deliver power to electric loads, and increased output of the FC. This results in an increased thermal output from FC, and heating from the boiler is lowered. The FC does not produce power exactly equal to its maximum generation capacity as discussed in Case 2 Section 5.2.

The utility tariff is at plain price during 13:00–16:00. This interval needs special attention as the proposed optimization model yields interesting results during this time interval. The BESS has options of either charging or discharging. Charging 1 kW at this interval with the 0.9 BESS efficiency results in 0.9 kW of discharge-able energy at the rate of 1 p.u./kW (0.9 p.u./0.9 kW). This rate is similar to the utility price of 1 p.u. at the interval 17:00–22:00. Charging in this interval does not result in any cost benefit for the system, and RCGA manages not to charge the BESS. On the other hand, discharging during this interval can result in cost reduction. However, the optimization algorithm weighs comparatively either to discharge during this interval or during 17:00–22:00. Discharging during a later interval saves more, therefore the BESS does not discharge during 13:00–16:00. In this way, the overall economy is optimized. Moreover, the electrical as well as thermal output of the FC is decreased to get benefit from the reduced tariff during this interval as shown in Figure 11.

Due to the peak price tariff during 17:00–22:00, a minimum energy is purchased from the utility. The BESS discharges and the FC increases its output to meet the energy demand. In the hours 23:00–24:00, the SH's electrical and thermal powers follow the trends already discussed in the interval 1 to 8.

The net cost of operation is 9.39($/day) in this case according to (9) which is less than the cost obtained in Case 5. Similar to Case 5, the scheduled charging of EV is considered in this case as well. It is notable that EV charging pattern in Figure 11 is different to that given in Figure 10. This is due to the fact that RCGA generates new random population for each simulation. No effort is made to reserve the randomness of the simulation process. Nevertheless, the EV is getting charged in valley hours in both cases and reaches to the SOC_{EVmax} before leaving the SH.

Table 4 shows the power demands of the electrical and thermal loads, and optimal powers from all resources for Case 6. The cost of the thermal and electric powers are for 24 h and are also shown in Table 4. Figure 13 indicates the convergence of cost function formulated in Equation (9).

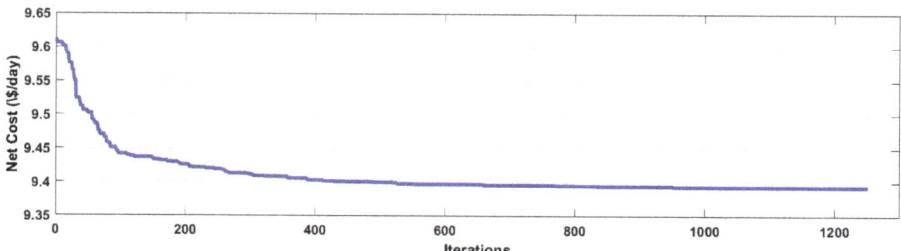

Figure 13. Convergence of RCGA.

Table 4. Powers and Costs.

	Powers (Demand and Generation)							Costs			
T_I h	P_{D_e} (kW)	P_{FC_e} (kW)	P_{EV} (kW)	P_{BT} (kW)	P_U (kW)	P_{D_h} (kW)	P_{FC_h} (kW)	P_{BL} (kW)	C_{FC} ($/day)	C_{BL} ($/day)	C_U ($/day)	Total ($/day)
1	1.55	0.91	1.35	−0.15	2.16	2.45	0.66	1.79	0.09	0.12	0.22	0.42
2	1.51	0.83	1.77	−0.25	2.72	2.41	0.59	1.82	0.09	0.1	0.28	0.47
3	1.49	0.98	1.63	−0.53	2.70	2.38	0.73	1.65	0.08	0.13	0.27	0.48
4	1.50	1.03	3.16	−0.35	4.01	2.34	0.77	1.57	0.08	0.13	0.41	0.62
5	1.53	1.02	1.53	−0.59	2.67	2.30	0.76	1.54	0.08	0.13	0.27	0.48
6	1.66	1.03	3.30	−0.25	4.20	2.28	0.76	1.51	0.08	0.13	0.43	0.64
7	1.91	1.09	1.75	−0.26	2.85	2.25	0.82	1.43	0.07	0.14	0.29	0.5
8	2.15	1.08	0.00	−0.62	1.74	2.29	0.81	1.48	0.07	0.14	0.18	0.39
9	2.30	1.62	0.00	0.65	0.05	2.30	1.39	0.91	0.05	0.23	0.01	0.28
10	2.38	1.74	0.00	0.42	0.23	1.95	1.56	0.39	0.02	0.25	0.03	0.3
11	2.40	1.61	0.00	0.68	0.12	1.98	1.38	0.60	0.03	0.23	0.02	0.27
12	2.35	1.54	0.00	0.58	0.25	2.15	1.28	0.87	0.04	0.21	0.03	0.29
13	2.33	1.58	0.00	−0.03	0.77	2.20	1.34	0.86	0.04	0.22	0.09	0.35
14	2.30	1.56	0.00	0.01	0.73	2.23	1.31	0.91	0.05	0.22	0.09	0.35
15	2.28	1.54	0.00	0.04	0.70	2.23	1.28	0.95	0.05	0.21	0.08	0.34
16	2.31	1.56	0.00	0.02	0.73	2.23	1.31	0.91	0.05	0.22	0.08	0.35
17	2.50	1.68	0.00	0.63	0.21	2.23	1.47	0.76	0.04	0.24	0.03	0.3
18	2.48	1.68	0.00	0.00	0.80	2.24	1.46	0.78	0.04	0.24	0.1	0.38
19	2.44	1.69	0.00	0.00	0.74	2.26	1.49	0.78	0.04	0.24	0.1	0.38
20	2.30	1.78	0.00	0.00	0.52	2.29	1.62	0.67	0.03	0.26	0.07	0.36
21	2.28	1.79	0.00	0.00	0.49	2.40	1.62	0.78	0.04	0.26	0.06	0.36
22	2.13	1.74	0.00	0.00	0.39	2.45	1.55	0.90	0.05	0.25	0.05	0.35
23	1.93	0.96	1.00	0.00	1.97	2.50	0.70	1.80	0.09	0.12	0.2	0.41
24	1.75	0.92	0.00	0.00	0.83	2.45	0.67	1.78	0.09	0.12	0.08	0.29

6. Conclusions

The modern SHs are foreseen to have an increased use of micro-CHP systems and availability of hybrid energy resources. This work presented a model of an SH and provided an algorithm for optimal scheduling of hybrid energy resources to minimize the cost of 24 h energy consumption. The findings of six different simulation scenarios reveal that the micro-CHP systems and the responsive electrical loads can play a vital role in reduction of the total energy cost. The conditions for the feasible use of BESS are also explained. The proposed optimization model based on successfully convergent RCGA makes use of the variable tariff and manipulates the devices for an optimal energy cost under the provided constraints. The presented work provides a comprehensive structure for hybrid energy management of a SH and can serve as a basis for further research. Further work can be carried out using a bidirectional utility grid, including thermal energy storage systems and integration of renewable energy resources.

Author Contributions: M.K.R. conceived the idea and did preliminary work. C.-H.K. provided guidelines necessary to complete the work. Z.M.H., M.I., and K.K.M. helped in code development. M.S.U.Z. and S.U.K. helped during paper writing and reviewing process.

Funding: This research was funded by National Research Foundation of Korea (NRF) grant funded by the Korea government (MSIP), grant number 2018R1A2A1A05078680" and "The APC was funded by the Authors".

Conflicts of Interest: The authors declare no conflict of interest.

References

1. Combined Heat and Power(CHP). Available online: http://aceee.org/topics/combined-heat-and-power-chp (accessed on 25 December 2016).
2. Liu, L.; Liu, Y.; Wang, L.; Zomaya, A.; Hu, S. Economical and balanced energy usage in the smart home infrastructure: A tutorial and new results. *IEEE Trans. Emerg. Top. Comput.* **2015**, *3*, 556–570. [CrossRef]

3. Nyboer, J.; Groves, S.; Baylin-Stern, A. *A Review of Existing Cogeneration Facilities in Canada*; Technical Report; Canadian Industrial Energy End-Use Data and Analysis Centre, Simon Fraser University: Burnaby, BC, Canada, 2013.
4. Aki, H. The Penetration of Micro CHP in Residential Dwellings in Japan. In Proceedings of the 2007 IEEE Power Engineering Society General Meeting, Tampa, FL, USA, 24–28 June 2007; pp. 1–4.
5. Houwing, M.; Negenborn, R.R.; Schutter, B.D. Demand Response With Micro-CHP Systems. *Proc. IEEE* **2011**, *99*, 200–213. [CrossRef]
6. Karami, H.; Sanjari, M.J.; Hosseinian, S.H.; Gharehpetian, G.B. An Optimal Dispatch Algorithm for Managing Residential Distributed Energy Resources. *IEEE Trans. Smart Grid* **2014**, *5*, 2360–2367. [CrossRef]
7. Murugan, S.; Horák, B. A review of micro combined heat and power systems for residential applications. *Renew. Sustain. Energy Rev.* **2016**, *64*, 144–162. [CrossRef]
8. Benam, M.R.; Madani, S.S.; Alavi, S.M.; Ehsan, M. Optimal Configuration of the CHP System Using Stochastic Programming. *IEEE Trans. Power Deliv.* **2015**, *30*, 1048–1056. [CrossRef]
9. Xie, D.; Lu, Y.; Sun, J.; Gu, C.; Li, G. Optimal Operation of a Combined Heat and Power System Considering Real-time Energy Prices. *IEEE Access* **2016**, *4*, 3005–3015. [CrossRef]
10. Adam, A.; Fraga, E.S.; Brett, D.J. Options for residential building services design using fuel cell based micro-CHP and the potential for heat integration. *Appl. Energy* **2015**, *138*, 685–694. [CrossRef]
11. Feng, Z.-B.; Jin, H.-G. Part-load performance of CCHP with gas turbine and storage system. *Proc. CSEE* **2006**, *26*, 25–30.
12. El-Sharkh, M.Y.; Rahman, A.; Alam, M.S.; El-Keib, A.A. Thermal energy management of a CHP hybrid of wind and a grid-parallel PEM fuel cell power plant. In Proceedings of the 2009 IEEE/PES Power Systems Conference and Exposition, Washington, DC, USA, 15–18 March 2009; pp. 1–6.
13. Nehrir, M.H.; Wang, C. Hybrid Fuel Cell Based Energy System Case Studies. In *Modeling and Control of Fuel Cells: Distributed Generation Applications*; Wiley-IEEE Press: Hoboken, NJ, USA, 2009; pp. 219–264.
14. Romano, R.; Siano, P.; Acone, M.; Loia, V. Combined Operation of Electrical Loads, Air Conditioning and Photovoltaic-Battery Systems in Smart Houses. *Appl. Sci.* **2017**, *7*, 525. [CrossRef]
15. Angrisani, G.; Canelli, M.; Roselli, C.; Sasso, M. Integration between electric vehicle charging and micro-cogeneration system. *Energy Convers. Manag.* **2015**, *98*, 115–126. [CrossRef]
16. Wakui, T.; Wada, N.; Yokoyama, R. Energy-saving effect of a residential polymer electrolyte fuel cell cogeneration system combined with a plug-in hybrid electric vehicle. *Energy Convers. Manag.* **2014**, *77*, 40–51. [CrossRef]
17. Wakui, T.; Wada, N.; Yokoyama, R. Feasibility study on combined use of residential SOFC cogeneration system and plug-in hybrid electric vehicle from energy-saving viewpoint. *Energy Convers. Manag.* **2012**, *60*, 170–179. [CrossRef]
18. Ribberink, H.; Entchev, E. Exploring the potential synergy between micro-cogeneration and electric vehicle charging. *Appl. Therm. Eng.* **2014**, *71*, 677–685. [CrossRef]
19. Yao, L.; Damiran, Z.; Lim, W.H. Optimal Charging and Discharging Scheduling for Electric Vehicles in a Parking Station with Photovoltaic System and Energy Storage System. *Energies* **2017**, *10*, 550. [CrossRef]
20. Mohsenian-Rad, H.; Ghamkhari, M. Optimal Charging of Electric Vehicles With Uncertain Departure Times: A Closed-Form Solution. *IEEE Trans. Smart Grid* **2015**, *6*, 940–942. [CrossRef]
21. Cao, Y.; Tang, S.; Li, C.; Zhang, P.; Tan, Y.; Zhang, Z.; Li, J. An Optimized EV Charging Model Considering TOU Price and SOC Curve. *IEEE Trans. Smart Grid* **2012**, *3*, 388–393. [CrossRef]
22. Jin, C.; Sheng, X.; Ghosh, P. Energy efficient algorithms for Electric Vehicle charging with intermittent renewable energy sources. In Proceedings of the 2013 IEEE Power Energy Society General Meeting, Vancouver, BC, Canada, 21–25 July 2013; pp. 1–5. [CrossRef]
23. Bhatti, A.R.; Salam, Z. A rule-based energy management scheme for uninterrupted electric vehicles charging at constant price using photovoltaic-grid system. *Renew. Energy* **2018**, *125*, 384–400. [CrossRef]
24. Wu, X.; Hu, X.; Yin, X.; Moura, S. Stochastic Optimal Energy Management of Smart Home with PEV Energy Storage. *IEEE Trans. Smart Grid* **2016**, *9*, 2065–2075. [CrossRef]
25. García-Villalobos, J.; Zamora, I.; San Martín, J.I.; Asensio, F.J.; Aperribay, V. Plug-in electric vehicles in electric distribution networks: A review of smart charging approaches. *Renew. Sustain. Energy Rev.* **2014**, *38*, 717–731. [CrossRef]

26. Khan, S.U.; Mehmood, K.K.; Haider, Z.M.; Bukhari, S.B.A.; Lee, S.J.; Rafique, M.K.; Kim, C.H. Energy Management Scheme for an EV Smart Charger V2G/G2V Application with an EV Power Allocation Technique and Voltage Regulation. *Appl. Sci.* **2018**, *8*, 648. [CrossRef]
27. Dubey, A.; Santoso, S. Electric Vehicle Charging on Residential Distribution Systems: Impacts and Mitigations. *IEEE Access* **2015**, *3*, 1871–1893. [CrossRef]
28. Qayyum, F.; Naeem, M.; Khwaja, A.S.; Anpalagan, A.; Guan, L.; Venkatesh, B. Appliance scheduling optimization in smart home networks. *IEEE Access* **2015**, *3*, 2176–2190. [CrossRef]
29. Javaid, N.; Ahmed, A.; Iqbal, S.; Ashraf, M. Day Ahead Real Time Pricing and Critical Peak Pricing Based Power Scheduling for Smart Homes with Different Duty Cycles. *Energies* **2018**, *11*, 1464. [CrossRef]
30. Nedstack. *Product Specifications of XXL Stacks*; Nedstack: Arnhem, The Netherlands, 2014.
31. El-Sharkh, M.Y.; Tanrioven, M.; Rahman, A.; Alam, M.S. A study of cost-optimized operation of a grid-parallel PEM fuel cell power plant. *IEEE Trans. Power Syst.* **2006**, *21*, 1104–1114. [CrossRef]
32. Gunes, M.B. Investigation of a Fuel Cell Based Total Energy System for Residential Applications. Master's Thesis, University in Blacksburg, Blacksburg, VA, USA, 2001.
33. National Household Travel Survey. *Electric Vehicle Feasibility: Can EVs Take US Households to Where They Need to Go?* Technical Report; National Household Travel Survey: Washington, DC, USA, 2016.
34. National Household Travel Survey. *Summary of Travel Trends*; Technical Report; National Household Travel Survey, Department of Transportation: Washington, DC, USA, 2009.
35. Hu, Z.; Han, X.; Wen, Q. *Integrated Resource Strategic Planning and Power Demand-Side Management*; Power Systems; Springer: Berlin, Germany, 2013.
36. Gianfreda, A.; Grossi, L. Zonal price analysis of the Italian wholesale electricity market. In Proceedings of the 2009 6th International Conference on the European Energy Market, Leuven, Belgium, 27–29 May 2009; pp. 1–6. [CrossRef]
37. Michalewicz, Z. *Genetic Algorithms + Data Structures = Evolution Programs*, 3rd ed.; Springer: London, UK, 1996.
38. Damousis, I.G.; Bakirtzis, A.G.; Dokopoulos, P.S. Network-constrained economic dispatch using real-coded genetic algorithm. *IEEE Trans. Power Syst.* **2003**, *18*, 198–205. [CrossRef]
39. Kuri-Morales, A.F.; Gutiérrez-García, J. Penalty Function Methods for Constrained Optimization with Genetic Algorithms A Statistical Analysis. In *MICAI 2002: Advances in Artificial Intelligence, Proceedings of the Second Mexican International Conference on Artificial Intelligence Mérida, Yucatán, Mexico, 22–26 April 2002*; Coello Coello, C.A., de Albornoz, A., Sucar, L.E., Battistutti, O.C., Eds.; Springer: Berlin/Heidelberg, Germany, 2002; pp. 108–117.
40. Amjady, N.; Nasiri-Rad, H. Economic dispatch using an efficient real-coded genetic algorithm. *IET Gener. Transm. Distrib.* **2009**, *3*, 266–278. [CrossRef]
41. Linkevics, O.; Sauhats, A. Formulation of the objective function for economic dispatch optimisation of steam cycle CHP plants. In Proceedings of the 2005 IEEE Russia Power Tech, St. Petersburg, Russia, 27–30 June 2005; pp. 1–6. [CrossRef]
42. Mitsubihsi i-MiEV Specifications. Available online: https://www.mitsubishi-motors.com/en/showroom/i-miev/specifications/ (accessed on 20 December 2016).
43. Saeed Uz Zaman, M.; Bukhari, S.B.A.; Hazazi, K.M.; Haider, Z.M.; Haider, R.; Kim, C.H. Frequency Response Analysis of a Single-Area Power System with a Modified LFC Model Considering Demand Response and Virtual Inertia. *Energies* **2018**, *11*, 787. [CrossRef]
44. Haider, Z.M.; Mehmood, K.K.; Rafique, M.K.; Khan, S.U.; Soon-Jeong, L.; Chul-Hwan, K. Water-filling algorithm based approach for management of responsive residential loads. *J. Mod. Power Syst. Clean Energy* **2018**, *6*, 118–131. [CrossRef]

© 2018 by the authors. Licensee MDPI, Basel, Switzerland. This article is an open access article distributed under the terms and conditions of the Creative Commons Attribution (CC BY) license (http://creativecommons.org/licenses/by/4.0/).

Article

Energy Storage Systems for Shipboard Microgrids—A Review

Muhammad Umair Mutarraf *, Yacine Terriche, Kamran Ali Khan Niazi, Juan C. Vasquez and Josep M. Guerrero

Department of Energy Technology, Aalborg University, 9220 Aalborg, Denmark; yte@et.aau.dk (Y.T.); kkn@et.aau.dk (K.A.K.N.); juq@et.aau.dk (J.C.V.); joz@et.aau.dk (J.M.G.)
* Correspondence: mmu@et.aau.dk; Tel.: +45-91778118

Received: 5 November 2018; Accepted: 11 December 2018; Published: 14 December 2018

Abstract: In recent years, concerns about severe environmental pollution and fossil fuel consumption has grabbed attention in the transportation industry, particularly in marine vessels. Another key challenge in ships is the fluctuations caused by high dynamic loads. In order to have a higher reliability in shipboard power systems, presently more generators are kept online operating much below their efficient point. Hence, to improve the fuel efficiency of shipboard power systems, the minimum generator operation with N-1 safety can be considered as a simple solution, a tradeoff between fuel economy and reliability. It is based on the fact that the fewer the number of generators that are brought online, the more load is on each generator such that allowing the generators to run on better fuel efficiency region. In all-electric ships, the propulsion and service loads are integrated to a common network in order to attain improved fuel consumption with lesser emissions in contrast to traditional approaches where propulsion and service loads are fed by separate generators. In order to make the shipboard power system more reliable, integration of energy storage system (ESS) is found out to be an effective solution. Energy storage devices, which are currently being used in several applications consist of batteries, ultra-capacitor, flywheel, and fuel cell. Among the batteries, lithium-ion is one of the most used type battery in fully electric zero-emission ferries with the shorter route (around 5 to 10 km). Hybrid energy storage systems (HESSs) are one of the solutions, which can be implemented in high power/energy density applications. In this case, two or more energy storage devices can be hybridized to achieve the benefits from both of them, although it is still a challenge to apply presently such application by a single energy storage device. The aim of this paper is to review several types of energy storage devices that have been extensively used to improve the reliability, fuel consumption, dynamic behavior, and other shortcomings for shipboard power systems. Besides, a summary is conducted to address most of the applied technologies mentioned in the literature with the aim of highlighting the challenges of integrating the ESS in the shipboard microgrids.

Keywords: energy storage technologies; hybrid energy storage systems (HESSs); microgrids; shipboard power systems; power quality

1. Introduction

Electrification in commercial and military ships has been a trend in recent past in order to reduce emissions and to improve efficiency [1–4]. The International Marine Organization (IMO) in 2012 stated that global SO_x and NO_x emissions from entire shipping exhibits about 13% and 15% of global SO_x and NO_x respectively [5]. It further states that for international shipping total CO_2 emissions are around 796 million tons, which are approximately 2.2% of the global CO_2 emissions. The CO_2 emissions from ships all over the globe is found to be 2.6% of the global CO_2 emissions. Moreover, IMO predicts that by 2050, CO_2 emissions in case of international shipping could raise in between 50% to 250%.

IMO announced guidelines and regulations in Jan 2015 for Emission Controlled Areas (ECA) as a consequence of modifications applied in the International Convention of the Prevention of Pollution from ships [6]. The European Commission set forth a novel climate agreement (*the Paris Protocol*) with an elongated ambition of diminishing global emissions up to 60% by 2050 as compared to 2010 levels [7].

In past, cost of energy and environmental concerns were not of greater importance as of now in marine power systems. In order to save fuel and decrease emissions, several solutions have been proposed. For instance: substituting alternative fuels, exhaust gases after treatment, and using hybrid propulsions are the frequently applied approaches implemented to achieve environmental guidelines imposed by IMO. However, these solutions for reducing emissions (SO_x, NO_x, CO_2, etc.) are not fundamental. Therefore, novel concepts such as hybrid energy storage systems (HESS) should be investigated in the shipboard microgrids.

There is an enormous evolvement over past few decades in shipboard microgrids due to their complex power architecture, power electronics interface based high power sources and loads. Hence, modern shipboard microgrids have become almost similar to terrestrial islanded microgrids, but due to the presence of high dynamic loads, complex control, and power management further complex the shipboard microgrids compared to terrestrial microgrids. Traditional power system relies on radial structure, which is used to have a separate generation for service and propulsion loads. However, due to the development of power electronics-based devices, the use of common power systems for both propulsion and service loads have been growing in the past few decades. Figure 1 depicts a single line diagram of the evolution of shipboard power systems.

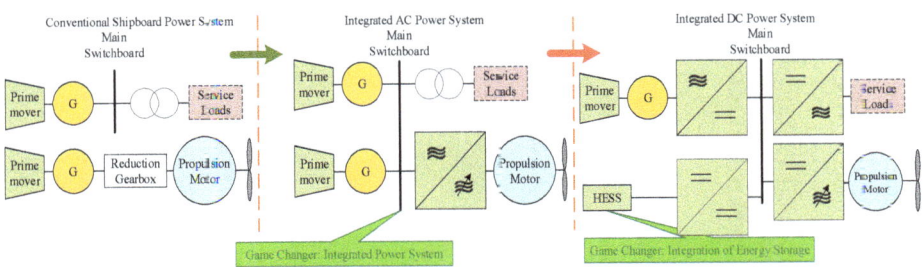

Figure 1. Evolution of Shipboard Power System.

The fast growth of electric and hybrid transportation systems open doors for further developments in ESS. Currently, the solutions are not technologically and commercially adequate in several features causing barriers to their broader usage. The ESS technologies vary from each other in terms of expense and technical aspects such as power density, energy density, charge and discharge time, operating temperature, lifetime, environmental impact, and maintenance requirement. Several works have been conducted in recent years, especially in the last decade, to improve ESS capacity. A typical single ESS technology, which can provide higher power and energy density, greater lifetime, and other such specifications, is not likely to be developed in near future. Therefore, in order to improve the capabilities, two or more ESS technologies can be hybridized.

Numerous modern technologies are being introduced in the maritime industry to meet the regulations imposed by various authorities. These technologies include liquid natural gas (LNG) as an alternative fuel, exhaust catalyst, hybrid propulsion, and so on. The implementation of Integrated Power Systems (IPS) have been firstly implemented in terrestrial Microgrid, then this application is extended to All Electric Ships (AES). The increased concerns over fuel economy and environmental issues have enforced maritime transport industry to hunt for fuel-efficient and lesser emission solutions. In marine vessels, power electronics offer a major role in fuel saving, particularly by the integration of ESS and electrification of propulsion systems through Variable Speed Drives (VSD). In order to address fluctuations caused by propulsion loads, several solutions have been proposed such as the use of thruster biasing for ships with dynamic positioning systems [8]. A thruster biasing is a situation in which thrusters on a ship start to act against each other, using more power than it is necessary to generate the commanded thrust. The thrust allocation algorithms such as presented in [9] bias the thrusters in such a manner that it consume a particular amount of surplus power. Later on, this surplus power can be released in order to prevent from blackout when the power generation capacity is reduced due to faults. This approach is applied in dynamic positioning systems and mainly suitable for low-frequency fluctuations. The other well-known solution is the integration of energy storage system to smooth the load power [10–12]. Using a single ESS technology may result in increasing the size, cost, and weight of the operated electric ships [13]. Therefore, HESS is found out to be a promising solution to cater to transients in shipboard power systems in an effective and efficient manner. Lately, mostly ESS is being used as an emergency power supply in the shipboard power systems. It can be helpful especially for offshore vessels in a dynamic positioning (DP) operation where the occurrence of faults may leads to the blackout, hence, in this scenario, ESS can power the propulsion systems for a shorter duration and can reposition the vessel during the fault until the ship is re-powered. Thus, there is a greater possibility of using HESS in future shipboard power systems as a power generation source for load levelling, peak shaving, and for reducing voltage and frequency deviations, which consequently may contribute in enhancing the power quality of the electrical power system. Figure 2 shows future shipboard zonal power system with an integration of HESS.

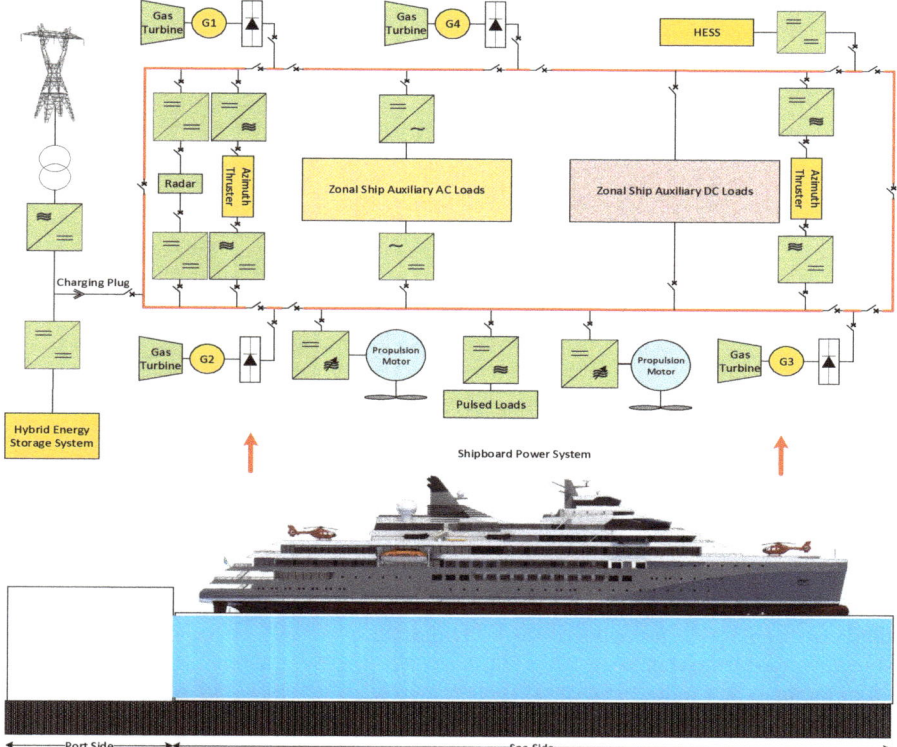

Figure 2. Typical single line diagram of the electrical power system for future ships.

The aim of this paper is to critically review

- Different types of energy storage devices such as batteries (lead-acid, Nickel Cadmium, Sodium Sulphur, Lithium ion), Ultra-capacitors, Flywheel, Superconducting Magnetic Energy Storage, and Fuel cells
- The energy storage devices that have already been used in marine vessels
- The most used hybrid combinations such as Battery-Ultracapacitors, Battery-SMES, Battery-Flywheel, and Battery-Fuel cell
- How the energy storage devices can enhance shipboard power systems
- What are the key challenges of integrating the ESS into the shipboard power systems

The rest of the paper is organized as follows: Section 2 presents an overview of different energy storage technologies. In Section 3, comprehensive analysis of the hybrid energy storage system are presented. Energy storage applications in shipboard power systems are discussed in Section 4. The challenges which occur while integrating an energy storage system in shipboard power systems are elaborated in Section 5. Finally, conclusion drawn from the study and authors opinion are presented in Section 6.

2. Energy Storage Technologies

An energy storage system comprises of an energy storage device, conversion of power and its control. Energy storage devices consist of secondary batteries, flywheels, capacitors, Superconducting Magnetic Energy Storage (SMES) systems, Fuel Cells (FCs), and pumped hydro. These devices differ from each other in terms of charge and discharge rate, life cycle, energy and power density, efficiency, etc. They are generally categorized into three groups with regard to the type of stored energy, i.e., electrical, chemical, and mechanical energy storage as shown in Figure 3.

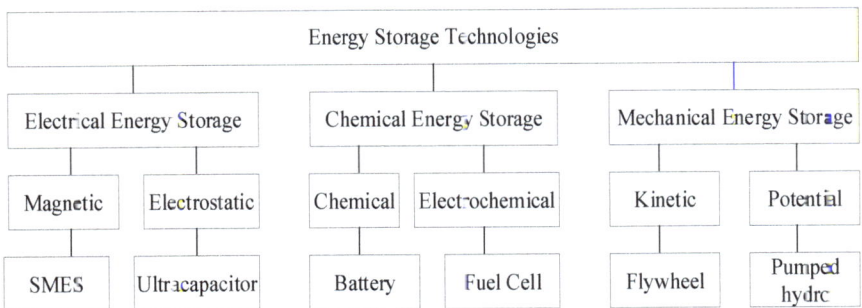

Figure 3. Energy Storage Technologies [14].

The maturity of different energy storage systems are depicted in Figure 4a, which are divided into three categories, mature, developed, and developing technologies. Lead-Acid is a mature technology, which has been used for over 100 years. NiCd, NaS, flywheel, ultra-capacitor, and SMES are the developed technologies that are commercially available. However, up till now, they are not used for large-scaled utility purposes. The fuel cell is still in the development phase as storing hydrogen is the key issue in this technology. The cycle efficiency of ESS is defined as $\eta = E_{out}/E_{in}$ where E_{out} and E_{in} is the output and input energy respectively. The efficiency of different energy storage technologies is depicted in Figure 4b, which shows that flywheel, SMES, and ultra-capacitor are highly efficient technologies.

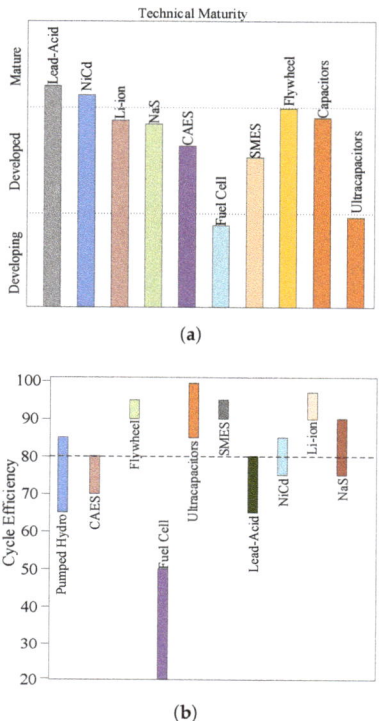

Figure 4. (**a**) Maturity and (**b**) Efficiency of different energy storage systems [15,16].

2.1. Batteries

Batteries are devices that transform chemical energy directly into electrical energy through an electrochemical oxidation-reduction reaction, and they are categorized as primary and secondary types of batteries. The former one cannot be charged electrically whereas the latter can. Lead acid battery is the most commonly used battery in the market. They are used in Uninterruptible power supply (UPS) [17–20], automobiles [21–23], etc. The increase in energy and power demands particularly, from hybrid electric vehicles results in large demand of batteries that are capable to produce higher energy density than lead-acid battery. The batteries that can provide an improved energy and power density are nickel metal hydride (NiMH) and lithium ion (Li-ion). Although their cost is high, still they are commercially adopted in various application, mainly in the automobile industry, cameras, medical instruments and in mobile phones.

Table 1 shows the different type of energy storage system with their power density, energy density, cost, efficiency, and lifetime, whereas Table 2 compares different type of energy storage technologies suitable for marine vessels. It can be seen that batteries such as lead acid, NiCd, NaS, and Li-ion are higher energy density devices, while flywheel, ultra-capacitor, and SMES are higher power density devices.

Table 1. Technical features of ESS [15,16,24,25].

System	Power Density (kW/kg)	Energy Density (kWh/kg)	$/kW	Efficiency (%)	Life Time (years)	Response Time
Lead Acid	75–300	30–50	300–600	65–80	3–15	ms
NiCd	150–300	50–75	500–1500	75–85	5–20	ms
NaS	150–230	150–240	1000–3000	75–90	10–15	ms
Li-ion	150–315	75–200	1200–4000	90–97	5–100	ms–s
Fuel Cells	500+	800–10,000	10,000+	20–50	10–30	ms–min
SMES	500–2000	0.5–5	200–300	90–95	20+	ms
Flywheel	400–1500	10–30	250–350	90–95	15–20	ms–s
Ultra-capacitor	100,000+	20+	100–300	85–98	4–12	ms

Table 2. Comparison between different battery technologies.

Type of Battery	Advantages	Disadvantages
Lead Acid	Inexpensive Lead is easily recyclable low self-discharge (2–5% per month)	Shorty cycle-life (around 1500 cycles) Cycle life is affected by depth of charge Low energy density (about 30–50 kWh/kg)
Nickel Cadmium	High energy density (50–75 kWh/kg) High cycle count (1500–3000 cycles)	High degradation High cost Toxicity of cadmium metal
Sodium Sulphur	High energy density (150–240 kWh/kg) No self-discharge No degradation for deep charge High efficiency (75–90%)	Temperature of battery is kept between 300 °C to 350 °C
Lithium-ion	Very high efficiency (90–97%) Very low self-discharge (1–3% per month) Low maintenance	Very high cost Life cycle reduces by deep discharge Need special overcharge protection circuit

2.1.1. Lead Acid

Lead acid batteries are the most used batteries in the world since 1890s [26] and are still extensively used in cost-sensitive applications where limited life cycle and less energy density are not of greater concern [27]. Their application includes stand-alone system with photovoltaic (PV) [28], emergency power supply system [29], mitigating output fluctuations from wind power systems [30], and as a starter batteries in transportation such as in vehicles [31]. They have small daily self discharge rate, typically less than 0.3%, fast response time, low capital cost, and relatively high cycle efficiency. The cycle life is around 1500 cycles at 80% discharge depth and the efficiency ranges between 80 to 90% . Furthermore, lead-acid battery is a mature technology, available at lower cost, easy recyclability, and simpler charging technique [32]. However, the drawbacks of this type of battery lies in lower energy density and using lead (a hazardous material). Moreover, it is not suitable for discharges over 20% of its rated value as it further reduces the life cycle.

2.1.2. Nickel Cadmium

Nickel Cadmium (NiCd) batteries have been commercially in use since 1915s. The battery uses metallic cadmium at the negative electrode and nickel oxyhydroxide at the positive electrode. It has a greater number of cycles, higher power and energy density as compared to lead-acid batteries. The lifetime of NiCd batteries at deep discharge range from 1500 to 3000 cycles depends on the type of the used plate [33]. This type of batteries are featured by the ability of working even at a lower temperature ranging from −20 °C to −40 °C. Moreover, these batteries are currently implemented only in stationary applications, which is prohibited in Europe on consumer use due to the toxicity of Cadmium and higher cost [34]. The best performance is achieved when discharged between 20% to 50% of the rated value [35].

2.1.3. Sodium Sulphur

Sodium Sulphur (NaS) batteries consist of liquid sodium at the negative electrode and liquid sulphur at the positive electrode, in between these two materials there is beta aluminium tube acting as an electrolyte. The cycle life of NaS batteries is 4500 cycles which is a bit higher than lead-acid batteries and the efficiency is around 75%. It is being particularly used in Japan over 200 sites for peak shaving. The temperature of this battery is kept in range of 300 °C to 350 °C. In order to maintain this temperature within that range, a heat source is needed so that their performance can be improved using its own mechanism, which results in affecting their performance.

2.1.4. Lithium Ion

Lithium-ion batteries in recent times have been of greater importance since the start of 2000, particularly in the area of mobile and portable applications such as laptops, cell phones, and electric cars. It has been proved that these batteries have exceptional performance especially in medical devices and portable electronics [36]. The nominal voltage level of each cell is around 3.7 volts as compared to 1.2 volts in the case of NiCd batteries. Another advantage is its higher energy and power density as compared to NiCd and lead-acid batteries. The main obstacle in using it, is the high cost, which is more than 600 $/kWh due to the overcharge protection and its specific packaging. Moreover, the efficiency of these batteries are quite high usually in the range of 95–98%, and the cycle life is around 5000. Safety is another severe issue in Li-ion batteries as most of metal oxide electrodes are unstable and may decompose at elevated temperature. Hence, in order to cater to this situation, the batteries are equipped with a monetizing unit such that to avoid over-discharging and over-charging.

2.2. Ultra-Capacitors

Capacitors store energy in terms of an electric field and generally known for their high symmetrical charge and discharge rates. Usually, capacitors have a quite low equivalent series resistances that enable them to supply the power efficiently. They are generally used in those applications where higher power is required for the shorter duration of time. The applications include camera flashes, filters, and compensation of reactive power. Capacitors are generally categorized by their dielectric medium, electrode material. They are further categorized as super-capacitors (also known as ultra-capacitors), electrolytic capacitors, and electrostatic capacitors. Figure 5 illustrates the individual structure of an ultra-capacitor. The key characteristics of ultra-capacitors are higher power density, faster charging and discharging due to lower internal resistance, enhanced life cycle, low voltage, and higher cost per Watt-hour (up to 20 times compared to Li-ion batteries). One of the main drawbacks of these ESSs is high sensitivity to over-voltage and, thus, overcharging. The other drawbacks include relatively low energy density, linear discharge voltage, high self discharge, and low cell voltage.

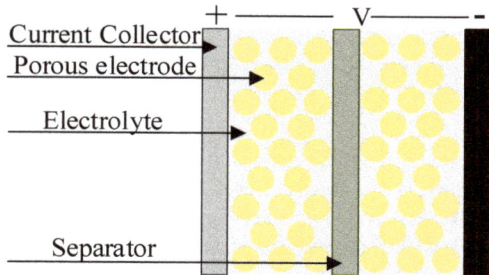

Figure 5. Individual Ultra-capacitor cell.

The life cycle of the battery is quite low in terms of charging and discharging cycles. Hence, in order to increase the lifetime of the battery and in particular to preserve system voltage above the minimum threshold, ultra-capacitors are hybridized with batteries in hybrid vehicles [37–39]. The use of hybrid electric vehicles comprises of batteries and ultra-capacitors are suggested in [40,41] and a commercially available ultra-capacitor based electric bus developed by Sinautec Automobile technologies [42], the range is around 5.5 miles. The studies have proved that by hybridizing battery and ultra-capacitor results in improving the lifetime, performance, and cycle life of the battery for hybrid vehicles.

2.3. Flywheel

Flywheel stores energy in terms of kinetic energy in rotating mass or rotor. The measure of energy stored depends on rotor mass, location of the mass, and rotor's rotational speed. In case of a certain amount of energy is stored in a flywheel, this could lead to an accelerating torque, which consequently results a flywheel to speed up. Moreover, when the energy is provided it could lead to decelerating torque, which might results in slowing down the flywheel. The energy stored E in a high-speed flywheel is given by:

$$E = \frac{1}{2}I\omega^2 \qquad (1)$$

where $I = \int r^2 \, dm$ (kg·m^2) denotes the moment of inertia of flywheel rotor and ω (rad/s) is the angular speed of the flywheel. The basic layout of flywheel is shown in Figure 6. In order to transform rotational kinetic energy to electrical energy, a flywheel must include a generator and motor. Likewise capacitors, the flywheel may have charge and discharge rates equal. They can be useful in improving power quality, peak shaving, power factor correction, and load leveling. Flywheels have been used widely in different applications such as UPS [43], frequency response [44], smoothing wind power [45], and heavy haul locomotives [46]. As compared to ultra-capacitors, flywheel provides intermediate characteristics in terms of power and energy density. Flywheel technology caters with many shortcomings of prior energy storage technologies by having limited temperature sensitivity, chemical hazardless, the similar rate of charge and discharge cycle, higher life cycle, reduced space, and weight.

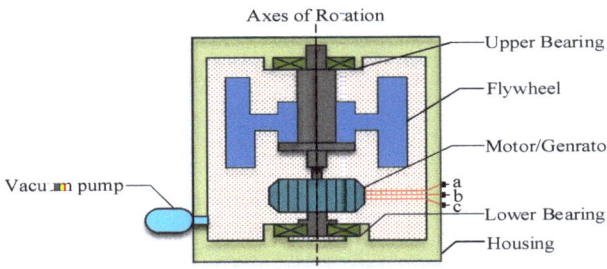

Figure 6. Basic Layout of Flywheel.

The study in [47] investigated and developed Flywheel energy storage system (FESS) for shipboard zonal power system. The main aim was to know where ESS can improve operation and/or reduce the maintenance cost. The applications where ESS can be beneficial includes "dark" start capability, system stability, pulse weapons, uninterruptible power supply, and load levelling. J. McGroarty et al. focuses "dark" start capability as an application of FESS in order to provide enough power capability and start opportunities to allow and help a gas turbine engine to come online from an off state. The optimization model for optimal sizing of FESS and dispatching controllable units economically for a drill-ship power system is presented in [48]. An optimization model of power management is proposed such that the optimization cost of vessel is minimized considering operational and technical parameters should not be violated. The proposed method further addresses how much flywheel energy storage system required to be installed and scheduling of various power plants considering several mission profiles and loading levels.

In the future, due to the increase size of all-electric ships there will be large amount of power sharing among different high power loads. In order to evaluate it, a model of a power train has been developed and is implemented in [49] for an electric ship. By using this model, the behavior of rotating machine power source have been explored in three different ways for a shipboard rail-launcher. Firstly, the impact of rapid charging of rotating machine on the shipboard power system is discussed by charging the rail launcher through 5 MW motors. Due to this, there would be a voltage sag that can be managed using stored energy in rotating machine (conceptually a FESS) to an appropriate level. Secondly, stored energy in the rotating machine is then used to improve the power quality of shipboard power system. In this study, by using an appropriate power electronics, the stored energy in the rail launcher can be used to correct the power quality issues introduced by rest of ship's power system. Finally, the energy stored in rotor of an alternator can be used to power a free electron laser for ship's defense. The rail-gun's power system is shown in Figure 7; the prime mover for the system is mainly fed from the ship's main power grid. The drive motor uses the power from the power grid in order to accelerate the rotor, which results into an energy storage in the rotating mass. Conceptually at this instant, system is having active components similar to a FESS. The stored energy is then used to launch the rail gun instead of using it elsewhere in ship. A capacitor is then discharged via rotor winding and to bootstrap the system to its full power, the induced current in the stator is fed back to the rotor, it takes around 30 ms. After achieving full power, the alternator launches the rail-gun. The pulsed alternator systems have low impendence field windings that rely on positive feedback self excitation or boot-strapping action in order to energize the field winding. So, a capacitor is used for the self-excitation process, which is discharged directly into the field winding. Furthermore, the field initiation capacitor is recharged through ship's main power system by a bi-directional converter.

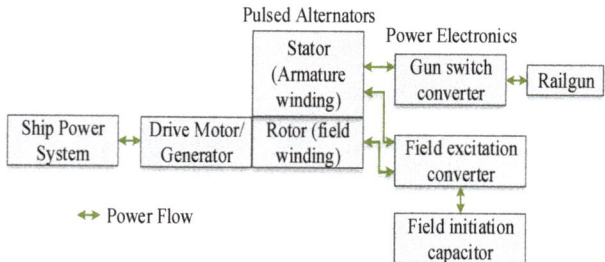

Figure 7. Basic block diagram of electromagnetic rail gun launcher.

The study in [50] addresses the use of high power FESS for DDG51 Arleigh Burke class destroyers to deal with high-power loads and to minimize the consumption of fuel. In the case of failure on one generator side, energy storage is responsible to power the critical loads up till another generator starts. The proposed study can mitigate transients in the system and provides a ride-through up to 10 min in order to start the backup generator. The study in [51] simulated FESS on the electrical power system based on offshore plants, which contains DP system such that to prevent from blackout, improve fuel efficiency, and mitigate voltage sags that usually take place in case of fault or the pulse loads. In the scenario, when there is an outage event of generator failure, the FESS will provide power until a backup generator starts. There are some particular rules specified for DP class, i.e., when the generator is operating at no load condition, at a nominal voltage level, and suddenly there is an additional loading, in this scenario the instantaneous voltage drop across the terminals of the generator should not be more than 15% of the nominal voltage. Furthermore, variations in the frequency should not exceed ±10% of the rated frequency and must be recovered within 5 seconds when the step load is turned off.

It is observed that by the use of FESS, the power system will overcome frequency drop and voltage sag within the limits to refrain from tripping other generators or blackout of power system. In [52], FESS is applied on the electrical network at the shipyard for powering the vessels from the shore distribution system such that to minimize fuel consumption on engines, avoid from blackout, and mitigate voltage sags. The simulation results show that there is around 15% drop in the rated voltage by the start of 2.25 HP motor in case when FESS is not integrated. On the other hand, by integrating FESS the voltage drop reduced to 4%.

2.4. Superconducting Magnetic Energy Storage (SMES)

Energy stored in SMES is in the form of a magnetic field created by superconducting coil. Initially, Ferrier introduced it in 1969 and originally it was anticipated as a load-levelling device [53] It is an energy storage method based on the fact that current will remain flowing through a superconductor even after the removal of voltage across it because of zero resistance [54]. In order to have negligible or zero resistance, the superconducting coil is sustained below the critical superconducting temperature with the use of an external cooling pump. The stored energy in the superconducting coil can then be released by discharging the coil. SMES storage devices are found to be highly efficient, i.e., greater than 95% as compared to other energy storage devices. The power electronic interfaces are needed that produces 2–3% loss in either direction. It has tiny deterioration because of cycling however, it has a high rate of self-discharge because of mechanical stability issues and energy spent on cooling it with cryogenic liquid. The magnetic energy stored in a conducting coil is given by:

$$E = \frac{1}{2}LI^2 \qquad (2)$$

where L is the inductance and I is the current. SMES system consist of three main components that are a super conducting coil, a cryostat system, and a power conversion system [55] as shown in Figure 8. Us Navy is trying to pull out from a stage dominated with hydraulic, pneumatic, and mechanical-based devices to a stage governed by electromechanical-based devices and with full electric control [56]. Future naval electric weapons require higher power pulses of electrical energy [57]. It is predicted that 200 MJ pulse forming network is necessary for the Navy's railgun to attain the anticipated muzzle energy of 63 MJ^2 [58]. SMES is found to be an attractive technology for this application, as it exhibits high energy density, zero resistance, and an efficient stored medium. SMES can further be helpful in providing power for onboard submarines, ships, and for naval applications [59].

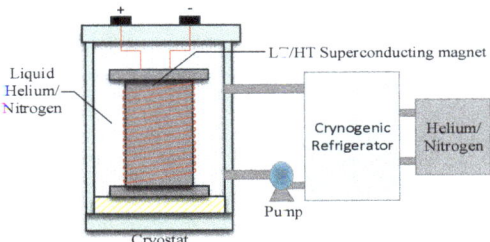

Figure 8. Superconducting magnetic energy storage (SMES) system.

2.5. Fuel Cell

Fuel cell transforms chemical energy directly into electrical energy and have the capability to be an alternative technology to the diesel engine, the individual structure of fuel cell is shown in Figure 9a. It has been proved to be more efficient as it produces lower or zero-emissions and functions cleaner as compared to a traditional gas turbine and an internal combustion engine. Polymer exchange membrane (PEM) fuel cell has been used to power Howaldtswerke-Deutsche Werft (HDW) submarines.

Nine Siemens PEM-based fuel cells were installed for propulsion purposes [60] ranging from 30–40 kW each. The first passenger ship to use fuel cell based propulsion is FCS Alsterwasser. The goal of the project was to test a ship that is free from emissions and to encourage the use of it for maritime applications [61]. The storage and hydrogen fuel distribution are the main challenging features for its wider use. As there are severe challenges to store hydrogen at a comparable energy density to hydrocarbon fuels such as liquid natural gas (LNG) or Marine Diesel Oil (MDO) [62]. It was suggested by Carlton et al. [63] that the technologies such as Molten Carbonate Fuel Cell (MCFC) and Solid Oxide Fuel Cell (SOFC) will be more favorable for ship propulsion as they use hydrocarbon fuels. The US energy department enlists several types of fuel cell [64] technology and are categorized as depicted in Figure 9b.

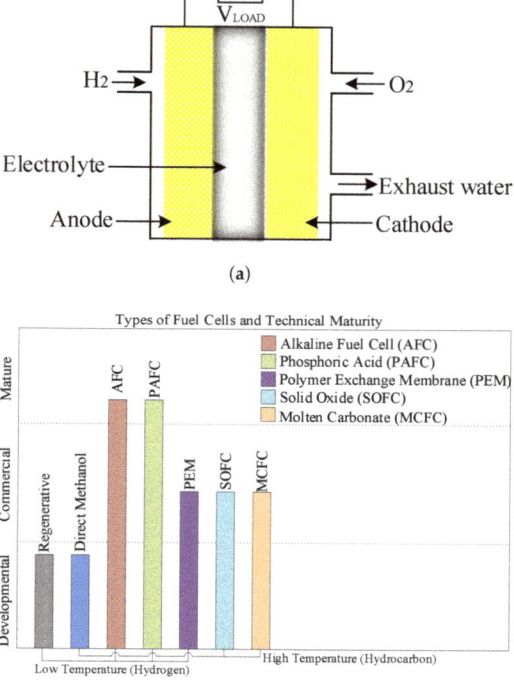

Figure 9. (a) Individual structure of Fuel cell and (b) Different types of fuel cell and their technical maturity.

Fuel cell for low emission ships (FellowSHIP), a research and development project with an involvement of industrial partners that comprises of Det Norske Veritas (DNV) (for the classification rules), Wärtsilä (for the energy), and Eidesvik Offshore (ship provider). The project is funded by Research council of Norway and its main goal was to integrate FC on offshore platforms and on-board vessels. In this research based project, a 330 kW FC is integrated with the *Viking Lady* as exhibited in Figure 10, an offshore supply vessel (OSV), the only commercially available vessel that uses fuel cell technology. It was docked in Copenhagen at the end of 2009 in order to replace traditional machinery to integrate fuel cell technology. The vessel is powered with dual fuel, i.e., liquified natural gas (LNG) and diesel-electric power plant. Four Wartsila based diesel engines and four main generators are installed to power the propulsion system and service loads. The vessel further uses molten carbonated based fuel cell and LNG to meet all the power needs. The Molten Carbonate Fuel Cell (MCFC) generates approximately 320 kW power and is operated around 650 °C. Hence, the combined use of gas engine and fuel cell results in the reduction of nitrogen oxide, sulphur oxide, and carbon dioxide emissions [65], and the efficiency of FC generated electric power was found to be 52.1% [66]. The concept study based projects that used fuel cell in the shipboard microgrids is enlisted in Table 3.

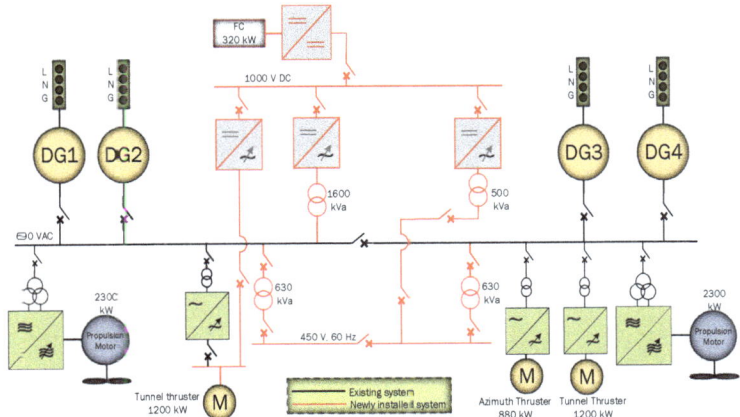

Figure 10. Fuel cell integration in *Viking Lady* [67].

Table 3. Summary of Fuel Cell based vessels and projects.

Vessel's Name	Type of Cell Used	Power Rating	Fuel Type	Reference
Viking Lady	MCFC	330 kW	LNG	[67]
Nemo H2	PEM	60–70 kW	Battery	[68]
ZemShip-Alsterwasser	PEM	100 kW	Battery	[69]
SF-BREEZE (Concept study)	PEM	2.5 MW	Liquid Hydrogen	[70]
PA-X-ELL (Concept study)	PEM	30 kW	Methanol	[71]
MV Undine (METHAPU Project)	SOFC	250 kW	Methanol	[72]
US SSFC (US Navy)	MCFC & PEM	2.5 MW	Diesel	[73]
MC-WAP (Concept study)	MCFC	500 kW	Diesel	[74]
MS Forester (SchIBZ Project)	SOFC	100 kW	Diesel	[71,75]
212 submarine U31	PEM	9(30–40) kW	H_2/Methanol	[76]
212 submarine (U32-36)	PEM	240 kW	H_2/Methanol	[75]
S-80 Class Submarine)	PEM	300 kW	Bioethanol	[77]
US Vindicator	MCFC	4×625 kW	F-76	[78]

3. Hybrid Energy Storage System (HESS)

The Hybrid Energy Storage System (HESS) is a combination of dissimilar energy storage technologies that have different characteristics with regard to energy capacity, cycle life, charging and discharging rates, energy and power density, response rate, shelf life, and so on. Figure 11 depicts the comparison of energy density, power density and their cost ($/kW).

ESS technologies can be categorized further into higher energy and power technologies. Higher energy devices such as a battery, fuel cell, pumped hydro, and CAES can supply energy for the longer duration of time but their power is low. On the other hand, higher power devices such as a flywheel, super-capacitor, SMES, and higher power batteries can supply very high power but for a shorter duration of time. It is observed that battery technology can be employed in both categories due to their wide characteristics range. Hence, hybridization of higher energy density devices with higher power density device will yield to a better ESS. In this way, high-energy devices will provide long-term power needs, whereas higher power devices will cater with short duration but higher power needs. Based on the above discussion, the possible combinations, which are extensively used in literature for different applications are depicted in Table 4.

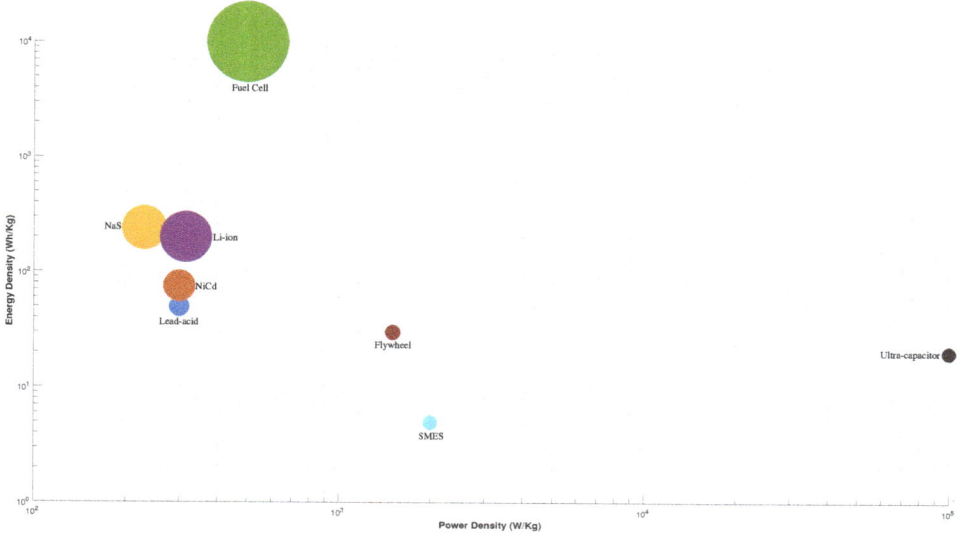

Figure 11. Ragone chart: Comparison of different energy storage technologies.

Table 4. Possible hybrid energy storage systems (HESS) configurations.

Energy Supplier	Power Supplier	References
Battery	Ultra-capacitor	[79,80]
	SMES	[81]
	Flywheel	[82,83]
Fuel cell	Flywheel	–
	SMES	[84]
	Ultra-capacitor	[85]
	Battery	[86]
CAES	Flywheel	[87]
	SMES	–
	Ultra-capacitor	[88]
	Battery	–

3.1. Battery-Ultracapacitor

Among all hybridized technologies, hybridization of ultra-capacitor and battery has been proposed in literature quite extensively. The rechargeable batteries are generally with high energy density and low power density normally below 1 kW/kg. The life cycle of the battery is quite low ranging from 1500 to 4500 cycles as compared to ultra-capacitors. In some literature, ultra-capacitors are named as supercapacitors and electrochemical double layer capacitors as well. This energy storage device has low energy density, typically below 10 kWh/kg and higher power density typically above 10 kW/kg. Furthermore, it possesses a high life cycle normally above 50,000. Generally, hybridization can be carried out via several methods, which can be categorized as internal and external hybridization [89]. In case of internal hybridization, the devices are developed by the hybridization of battery and ultra-capacitor on the electrode level as shown in Figure 12a. The hybrid battery pack system "UltraBattery" [90] is an example of internal hybridization as shown in Figure 12b. It is the hybridization of lead-acid battery and ultra-capacitor and was developed by Commonwealth Scientific and Industrial Research Organization (CISRO) in Australia. On the other hand, the hard wire connection between a ready available battery and ultra-capacitor is categorized as "external hybrid" as shown in Figure 12c. Among the methods mentioned above, the external hybrid method is the extensively used method in several applications.

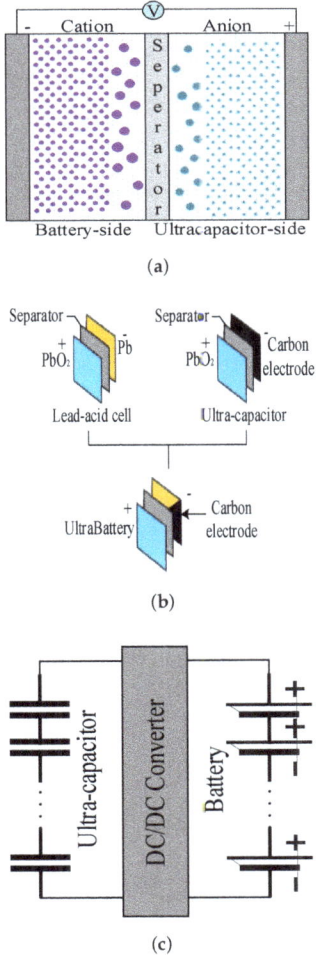

Figure 12. Possible approaches for hybridizing ultra-capacitors and batteries. (**a**) Internal hybridization. (**b**) Ultrabattery (**c**) External hybridization.

Yichao tang et al. [79] explores the feasibility by hybridizing battery and ultra-capacitor energy storage for naval applications. A dual active bridge based topology is proposed to control the bi-directional power flow through phase shifting for both charging and discharging ultra-capacitors and batteries. The topology is designed in such manner that it can meet the requirements of both 1 MW pulsed load and 100 to 500 kW propulsion system. Higher frequency switching devices were selected such that to achieve DC-DC conversion at higher power and voltage levels. The electric propulsion system in vessel experiences large torque and power fluctuations on their drive shaft because of waves and rotational motion of propeller. Jun Hou et al. [80] explores novel solutions to address such fluctuations by exploring energy management strategies and integration of ultra-capacitor and battery. The two main objectives, i.e., mitigation of power fluctuations and HESS loss minimization are assessed at different sea conditions. The simulation results depict that substantial benefits can be attained in terms of reduction in fluctuations and losses. During navigation of ships at sea, it suffers from a constant rocking motion and is affected by ship navigation parameters and surrounding sea conditions, which further increases the uncertainty involved with the use of solar energy in ships. In [91], a mathematical model is considered for generating power through PV modules while considering both the sea conditions and movement of the ship. The rocking motion of vessels fluctuates the power typically for 10–20 s and it can reduce the lifetime of the battery to quite an extent. Hence, to cater to situation hybridization of ultra-capacitor with lithium-ion battery is proposed to improve the stability and reliability of the power system. The hybrid energy storage system based on ZEBRA batteries and ultra-capacitors modules for All-electric ships were considered in [92] to decrease the battery charge and discharge peak currents. Ultra-capacitor modules were considered in order to extend the expected life of the battery. Cohen et al. [93] presented an actively controlled Li-ion battery hybridized with ultra-capacitor for pulsed power applications aiming to maximize the energy density of Li-ion battery and also to maximize the energy and power density of Ultra-capacitor. Furthermore, the authors designed, constructed, and validated the hybrid model using commercial off-the-shelf technologies and it is observed that the generator's frequency and voltage deviations are massively improved.

3.2. Battery-SMES

The capacitor's nominal voltage is quite low ranging from 1 to 2.5 V due to the fact that the series connection of numerous units is required in order to provide higher voltages. However, connecting several units together in series can cause voltage imbalance. So, in order to balance the voltage, some protection circuits are required, these interfaces further may cause fluctuations in the power system, hence, step-down and step-up converters are further installed to adjust the output voltage. The change in output voltage in ultra-capacitors varies with its charge and discharge and is proportional to the stored energy. In contrast to ultra-capacitors, superconducting magnetic energy storage (SMES) does not require any step-down or step-up converters. It is basically a superconducting coil that stores energy in the form of a magnetic field. It has the capability to deliver from/to power system with outstanding characteristics such as high efficiency, high power density, fast response time, and higher life cycle. The implementation of SMES system is difficult, as it requires the refrigeration mechanism that is quite costly and involves complex maintenance. The special site requirements further limit its application that is stationary such as railway supply substation and renewable generation sites. Some researchers who proposed and investigated battery-SMES-based HESS system for transportation applications are as follows.

The demand for all-electric ships (AESs) amplified rapidly in recent times and load fluctuations in the system may lead to severe issues such as increased fuel consumption, voltage fluctuation, and environmental emissions. HESS comprising of a battery (higher energy density) and SMES (higher power density) proposed in [81] in order to cater to shiploads that cause sudden changes such as maneuvering and pulse loads. As the ramp-rate of vessel-generators such as gas-generators usually are in between 30–50 MW/min range and on the other hand, pulsed load require 100 MW/s ramp-rate that is far higher than the ramp-rate of generators [94]. Hence, ESS has become vital to deliver a huge amount of energy within a short period.

3.3. Battery-Flywheel

Jun Hou et al. [82] proposed a hybrid battery and flywheel energy storage system in order to isolate load fluctuations from the shipboard power network. The effectiveness and feasibility of the proposed hybrid system to mitigate load fluctuations for all-electric ships under various sea conditions are shown through simulations. Li-ion battery was used due to their higher power and energy densities than other batteries.

In [83] the authors explored a novel solution by using flywheel and battery as a hybrid model in order to address fluctuations in load power. It is shown through simulations that with the use of battery-flywheel, the effectiveness and feasibility is quite high such that to mitigate load fluctuations, especially at high sea states. As power fluctuations may result into reduction in electrical efficiency, uncertain consumption of power, and most probably affect shipboard power quality. In [95] flywheels have been analyzed such that to address pulse power loads. The results depict that by using flywheel energy storage system, the stability of shipboard power system can be maintained during operation of pulse load. AT Elsayed et al. [96] presented a comparative study in order to determine the optimal hybridization of batteries, flywheel, and ultra-capacitors to minimize the frequency and voltage fluctuations, which are produced in a result of adding pulsed loads to the shipboard power system either on the AC or DC side. Hai Lan et al. [97] modeled a high-speed FESS in order to smooth the photovoltaic power fluctuations and hence improving the power quality of a large oil tanker. The sinusoidal pulse width modulation (SPWM) along with constant torque angle control method is proposed such that to control charging and discharging of a flywheel.

3.4. Battery-Fuel Cell

Hydrogen-based fuel cell presently has been of greater importance in the maritime industry which includes: Nemo H2, Hydrogenesis, Hydra, and fuel cell ship (FCS) Alsterwasser. Reduction in consumption of fuel, lesser emissions, negligible noise, lower maintenance requirements, and minimal vibration are the key features which led in developing maritime fuel cell technology. In 2008, Alster-Touristik GmBH developed FCS Alsterwasser, it was a first passenger tourist vessel that was entirely powered by fuel cells. FCS Alsterwasser can withstand up to 100 passengers operating at cruising speed of 8 knots, it has two 50 kW fuel cells powering 100 kW hybrid electric propulsion system in combination with lead-acid batteries [48]. Henderson further states that it is estimated that approximately 220 kg of SO_x, 77 tons of CO_2, and 1000 kg of NO_x is saved annually as compared to the traditional diesel-powered vessel. In December 2009, Nemo H2, a zero-emission canal boat was developed by Fuel Cell Boat B.V. It has the capacity of 87 passengers and 1 crew member operating at a cruise speed of 16 km/h. A hybrid propulsion system comprising of 60–70 kW PEM-based fuel cells with 30–50 kW batteries were installed. 24 kg of Hydrogen is stored in 6 cylinders at a pressure of 35 MPa [68].

The requirements of military submarines are quite severe, such as they longer underwater operation, low transfer of heat to sea water, low magnetic signatures, and low noise levels [98]. Traditional submarines are equipped with a diesel-electric based propulsion system and for underwater operations, battery energy is used (lead-acid). The batteries were charged using diesel generators during snorkelling period. Hence, fuel cells are found to be a possible alternative candidate in order to meet the specific requirements associated with air-independent propulsion (AIP) system [99]. In the 1980s, the German navy in collaboration with Siemens has tested 100 kW Alkaline Fuel Cell (AFC) system in an onshore laboratory and then in Class 205 submarine U1 [98] in order to judge the application of fuel cells for the submarine. The system consists of 16 × 6.2 kW Siemens modules of AFC, four modules each connected in series in order to correspond to battery's voltage and propulsion system. Later on, Siemens developed a 34 kW Polymer electrolyte fuel cell (PEFCs) module for German Class 212 submarines [100]. Submarine class 214 was launched in 2005, it uses two Siemens-based 120 kW PEFC modules [101]. It is connected to the main grid via DC to DC converter and has the efficiency of 56% on full load.

A hybrid fuel cell/diesel generator power system is proposed in [102] for propulsion and for test equipment on a research vessel. The PEM-based fuel cell system with a battery as backup and secondary energy source simulated in power system computer aided design (PSCAD). The secondary source is a lead-acid battery with a rating of 360 V and energy 82 kWh. The simulation-based analysis depicts that the system has the capability to handle sudden load changes with minimal transients. Although FC's are a promising solution to reduce greenhouse gas emissions but their response time is not fast enough to cater to load transients that might occur in vessels at sea. Hence, higher density secondary batteries are needed to accomplish stability under transients and usually, dc/dc converters are needed for interfacing battery and FC into the DC link. Alireza et al. [103] presented an intelligent power strategy in order to improve the performance of FC without utilizing dc/dc interfacing converters. A new FC power management based strategy by using genetic algorithm proposed in [103] such that to guarantee the efficient performance of FC stack by preserving FC voltage within a required range in FC–battery hybrid system without the use of DC/DC interfacing converters.

The study in [104] proposed a hybrid system based on battery and PEM-based fuel cell to control power generation in a shipboard power system. The mathematical model for regulating active and reactive power is derived and integrated with PEMFCs in order to enhance the system dynamic response. Test results illustrate that injunction of hydrogen fuel into the fuel cells can be regulated automatically with fluctuations in loads. Furthermore, the batteries are used to compensate power in order to maintain operational security of the system.

4. Energy Storage Applications in Shiborad Microgrids

ESS can provide benefits to marine vessels as follows:

- Improves the stability of the system, which arises due to slow response of the engines to load demand.
- Decreases operational cost due to less engine maintenance and by optimizing fuel consumption.
- Minimizes the risk of blackout by installing an ESS as a UPS, such that it provides quicker response to a blackout as compared to emergency generators.
- ESS acting as an additional power reserve, hence provides power in case of failure of a generator. Furthermore, it can minimize the number of generators that have been online to improve the redundancy of the power system.
- ESS can also be helpful in peak shaving, load levelling, power smoothing, frequency and voltage fluctuations, and power quality.
- Decreasing thruster load ramp limits by adding inertia through ESS, which limits the power slew-rate and enables quick thrust force. Therefore, it enables quick response of vessel and boosts the capabilities of maneuvering.

Figure 13 illustrates the discharge timings of different applications of the stored energy. It can be seen that load levelling, peak shaving, and power smoothing are applications that take long time to discharge the energy, which has been stored in the ESS. Therefore, Li-ion batteries can be used for these applications. On the other hand, UPS and black start applications require high power density, so ultra-capacitors and flywheel are the most recommended ESS for these purposes. Table 5 compares different applications of energy storage devices in shipboard microgrids.

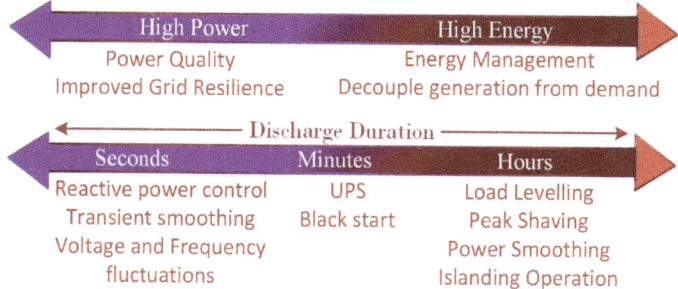

Figure 13. Energy storage applications and their discharge timings.

Table 5. Comparison between different applications of energy storage system (ESS).

Type	Functionality	Stored Capacity
Peak shaving	peak the shaves by energy storage or additional generators	Large
Power smoothing	smooth short term fluctuations by adding local energy storage system	Small
Power ramp-rate limitation	limit the power slew-rate by the addition of energy storage	Small
Load levelling	store energy to ESS when electricity is cheap or when there is light loading and delivers it when the electricity is expensive or when there is a high load demand	Medium

4.1. Load Leveling and Peak Shaving

Load levelling stores power when there is a light loading on the power system, then it delivers it during the period of high demand. When there is a high load demand, energy storage system supplies power and hence reduces the demand of load, which results in less economic peak generating facilities. On the other hand, moderate demands and reducing the peaks are called peak shaving. It is normally adopted at a higher scale by power companies in order to save money and sometimes by the commercial companies to sell the power and gain money, and thereby they purchase the power during low demand and sells it during high demand [105].

At present, ESS particularly Li-ion battery has been adopted to cater to variable loads in all-electric ships that are on shorter routes such as MF Ampere (route length 5.6 km), MF Folgefonn (route length 5.6 km), and Aero Ferry (route length up around 24 km). The cost-effective benefits are derived from peak shaving, spinning reserve, and load levelling functionalities [106]. Therefore, ESS can be useful in reducing the size and the number of generators to deal with variable loads by shaving the peaks or levelling the loads, which consequently reduces the emissions, fuel consumption, as well as wear and tear on the engines. In [107], distributed ESS contains a NaS battery that is utilized to shave the peak in order to mitigate the capacity constraints. In [108], 1 MW ESS based on Li-ion battery is installed at Nagasaki Shipyard for peak shaving operations. One prospect is to use ESS for peak shaving application as illustrated in Figure 14. When a marine vessel approaches near the harbor and is required with swift response in maneuvering, deprived of starting additional generators, ESS can be helpful in this scenario.

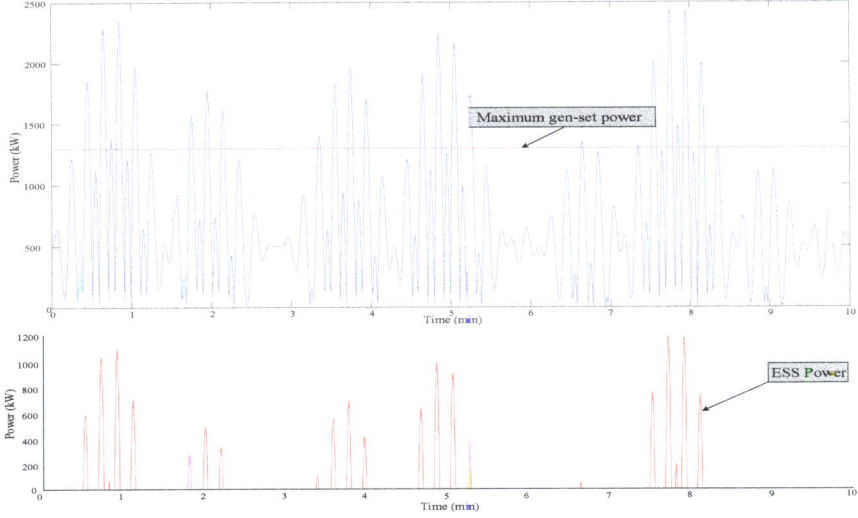

Figure 14. Peak shaving application in marine vessel.

The applications where lifting and lowering operations are required, for instance, in drillship and cranes there is a possibility to configure ESS in such a manner that it absorbs the regenerated energy instead of dissipating it in the dynamic braking resistors. This stored energy then can help to shave the peaks and in load levelling as shown in Figure 15.

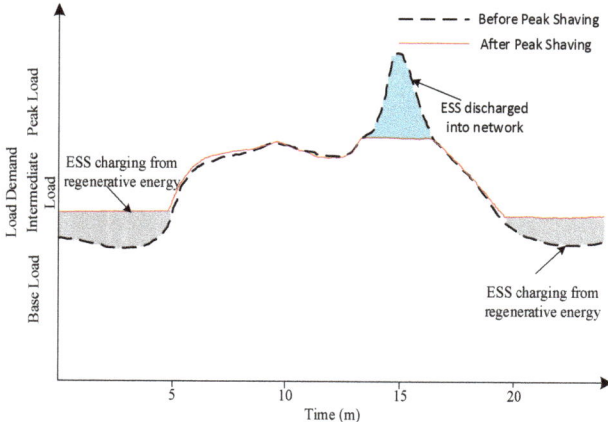

Figure 15. ESS application for peak shaving and load-levelling.

4.2. Power Smoothing

The battery-based power smoothing control in a shipboard microgrids based on using non-linear predictive control is proposed in [109]. In large vessels power fluctuations are quite high, which results in frequency fluctuations and can cause wear and tear of the source power plants. To cater to this issue, integrating batteries with DC/AC drive has been proposed by the same authors. However, due to the high fluctuations caused by the propulsion loads, which lead to an increase in temperature of the batteries, it is recommended to use a band pass filter with an optimized cutoff frequency parameters based on model predictive control. The energy storage systems such as batteries can also be added next to propeller motors to smooth the power oscillations as depicted in Figure 16. Power smoothing strategies have been utilized quite often in intermittent renewable energy such as in wind and solar energy conversion systems. Energy storage system can smooth the power by storing the energy from peaks and controls the ramp rate (MW/min) in order to eliminate rapid voltage and power fluctuations from the grid. Ultra-capacitor, fuel cell, battery, flywheel, and SMES are the energy storage technologies, which have been particularly used in wind energy for power smoothing applications. Therefore, these energy storage technologies can be helpful in smoothing electric power for shipboard microgrids as well.

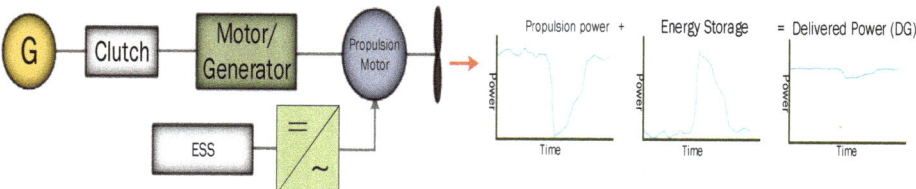

Figure 16. Power Smoothing utilizing energy storage technologies.

4.3. Frequency and Voltage Fluctuations

The heavy loads such as pulsed loads (propulsion motors, pumps, thrusters, etc.) can draw a large power in a short duration of time. If this amount of power exceeds certain limits, it might result in voltage and frequency fluctuations. In order, to avoid such voltage and frequency fluctuations, several standardization authorities have defined limitations for such fluctuations. Among these, IEC 60092-101 is an extensive standard, which defines the limitation for frequency and voltage fluctuations as depicted in Table 6. The other standards DNV [110] and Lloyds Register of Shipping (LRS) have the same limitations only for frequency, and for voltage fluctuations in case of a shorter period, they have set the limitation between −15% to 20%. In addition, in case of the steady state, DNV standards have limited the voltage fluctuations to ±2.5%.

Table 6. Frequency and voltage permissible level as per IEC 60092-101 in shipboard microgrids.

	IEC 60092-101		
	Steady State	Transient State	
	Magnitude	Magnitude	Duration
Voltage	−10% to +6%	±20%	1.5 s
Frequency	±5%	±10%	5 s

ESS plays an important role in enhancing voltage and frequency fluctuations. In [111], a HESS based on improved maximum power point tracking (MPPT)-based algorithm is presented to enhance the performance of photovoltaic plant that is installed in the shipboard power system. This strategy helped to smooth and regulate the frequency oscillations. Besides, frequency hierarchical-based control algorithm is utilized to assign lower frequency oscillations to the battery and higher frequency oscillations to the ultra-capacitor. The improved MPPT algorithm further helps in reducing the installed capacity of HESS. The effectiveness of the proposed approach is verified under shipboard power system model.

Viknash Shagar et al. [112] utilized advance control strategies such as model predictive control (MPC) to minimize frequency fluctuations within the permissible limit as recommended by power quality standards in a shipboard power system. Moreover, battery-based energy storage system is directly connected to the DC link of the frequency converter as compared to traditional approaches in which battery energy storage system (BESS) is connected via DC-DC converters, hence reduces the complexity of ESS. It is further observed that the changes in service loads have a less impact compared to the changes in propulsion loads, which have a higher impact on bus bar frequency. Also, BESS has been integrated into a shipboard power system to mitigate these frequency fluctuations. The study in [113] utilized 1000 kW BESS with a DC-link capacitor and an active front-end (AFE) converter in order to boost the voltage and frequency quality together to suppress the grid harmonics as depicted in Figure 17.

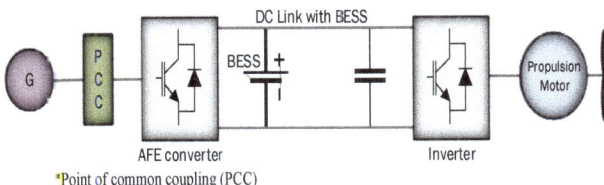

Figure 17. Integration of BESS with integrated power systems (IPS).

In order to verify the approach presented in [113], authors took a transient load that causes sudden change in the frequency of grid and ultimately trips the generator and pulse load (an active and reactive power load), which causes sudden change in voltage and frequency of the power system. The overall load of 8.17 MW comprises of a trapezoidal load that consumes 5 MW of power for 200 ms, service load that consumes 1 MW with 250 kVar, and the propulsion load that consumes 2.17 MW. It is observed that with BESS compensation strategy the deviation of voltage at point of common coupling (PCC) was decreased to less than 10%, and the deviation in frequency was reduced to 2.5% as well, which satisfies the standards.

4.4. Power Quality

Nowadays, power quality has become a hot topic in shipboard microgrids. These power quality issues can be voltage dips, voltage and frequency fluctuations, harmonic contamination, and flickers, etc. Harmonic distortion is one of the main issues, which arise due to the presence of non-linear loads of the electric power system. The extensive use of non-linear loads in shipboard microgrids as compared to the terrestrial power system is a major concern, as these loads draw non-linear current while flowing through the power system resulting in distortion in the waveform of voltage, hence affecting the whole power system. The use of power electronic based converters in vessels such as in propulsion system, compressors, and thrusters has increased to a high level, for example, in the non-linear loads in navy and cargo vessels can reach up to 80% of the overall onboard capacity. The frequency variations and high extent of inter-harmonic distortion in a shipboard power system make it difficult to measure harmonics. Most used solutions for power quality issues in the terrestrial grids is the application of passive and active power filters [114,115]. Since the shipboard microgrids can be considered as a self-sufficient microgrid, the solutions applied to the terrestrial grids can be extended to shipboard microgrids as well. In regard to IEC 61000-4-7 standard, the maximum allowable synchronization error between power system frequency and synchronization should be within a range of ±0.03% of the nominal frequency of power system. In order to cope with this problem, several methods for estimating harmonics in a shipboard power system with the use of Fourier analysis by considering different synchronization methods and sampling windows is described in [116]. Different marine classification bodies, which include DNV, LRS, IEEE Std 45-2002, IEC 60092-101, and American Bureau of Shipping (ABS) have proposed limitation for power quality issues of a shipboard microgrids. As stated in IEC 60092-101, the Total Harmonic Distortion (THD) should not exceed more than 5% limitation considering no single harmonic should be greater than 3% of the fundamental voltage. The voltage harmonic distortion limitation from different classification bodies are depicted in Table 7.

Table 7. Harmonic voltage distortion limitation in shipboard power system.

	Different Standards				
	DNV	LRS	IEEE Std 45-2002	IEC 60092-101	ABS
Total harmonic distortion (THD)	8%	8%	5%	5%	5%
Single harmonic distortion	5%	1.5%	3%	3%	3%

To improve the power quality of shipboard power system FESS is utilized in [117]. This topology can store energy upto 80 MJ. An induction motor is considered as a propulsion motor with a power rating of 20 MW and a high power pulse electrical equipment with power rating of 2 MW is integrated as shown in Figure 18. It is observed that it has a huge impact on the power quality during the start of the high power pulsed load. However, the integration of FESS results in reducing the frequency fluctuations. In [118], a series voltage injection of FESS is presented for mitigating voltage sags in a shipboard power system for maximizing ship's survivability. The scheme basically comprises of a power electronic interface, a flywheel energy storage system coupled with an induction machine, and series injection-based transformer. The stored energy in a FESS, therefore, helps to mitigate voltage sag problems, especially for critical loads. Modeling, simulations, and analysis of FESS interfaced with power converter using PSCAD/electromagnetic transients including DC (EMTDC) has been presented in the proposed approach. In [119] authors proposed multi-modular DC-DC converter-based FESS for shipboard power system, particularly for medium voltage direct current (MVDC) grid application. By using a virtual impedance based strategy, the cell capacitors of multi-modular converter are applied for implementing dc active power filter capabilities, which helps to improve the bus power quality without the use of auxiliary devices or even sacrificing battery lifetime. The study in [50] proposed FESS as a feasible approach to save significant fuel on the DDG51 Arleigh Burke class destroyers to support pulsed power based loads. The FESS allows the use of a single turbine generator close to a full load rather than using the traditional practice of running two turbine generators at less than a half load. In this strategy, it helps in saving the fuel, improves the reliability and power quality of the shipboard power system.

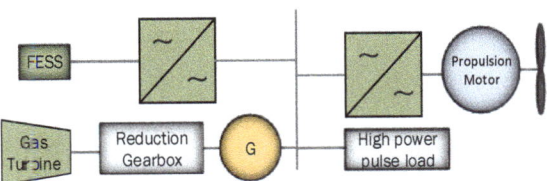

Figure 18. Integration of flywheel energy storage system (FESS) with IPS.

4.5. BESS Based Vessels

MF Ampere, the first zero-emission ferry operated by Norled AS, which sails on a 5.7 km crossing between the villages of Lavik and Oppedal, Norway. It has the capability to carry up to 120 cars and 360 passengers [120]. The vessel consists of two onboard motors, one of the motor is used to drive the thrusters, these motors are operated by lithium-polymer-based batteries [121]. The sailing time of the ferry is approximately 20 min and an extra 10 min are particularly specified for charging the batteries from the battery station located at shore side as shown in Figure 19. In order to recharge the batteries faster and not to put burden on the village grid, battery banks are installed on each side of shore to recharge the batteries. Approximately, one million litre of diesel per year is saved, 15 tons of nitrogen oxide and 570 tons of carbon dioxide emissions are reduced as compared to the same size of the vessels powered by traditional power sources operated on a similar route.

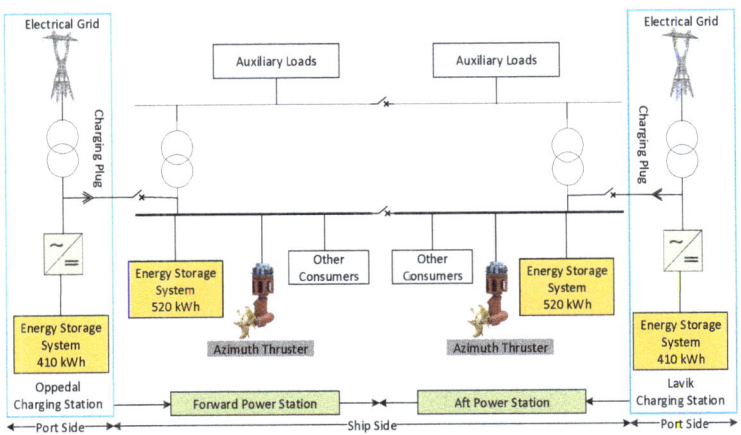

Figure 19. Schematic diagram of *MF Ampere*.

The first fully electric cargo ship was launched in Guangzhou by Guangzhou Shipyard International Company Ltd in 2017 to carry coal for thermal power generation along Pearl river. It is the first cargo ship to use Li-ion batteries with such a huge power rating of 2.4 MW and it takes approximately 2 h to charge these batteries [122]. The vessel carries up to 2000 tons of cargo can cruise up to 80 km when the batteries are fully charged at a maximum speed of 7 knots. The vessel is emission free and does not emit any greenhouse gases, hence contributing towards the pollution and fossil-fuel free future. *Greenline 33*, a hybrid yacht with diesel/electric propulsion system, the schematic diagram of this yacht is depicted in Figure 20. The yacht further consists of 6 solar panels, they are installed on the roof top of the yacht to charge the battery bank. Li-ion-based batteries are installed to run the yacht in an electric mode. It is observed that the yacht is emission free and the running cost in case of electric mode is reduced by 10 times as compared to the diesel engine. *Greenline 40* and *Greenline 46* are among the ongoing projects, which are based on fully electric propulsion system with Li-ion battery packs [123].

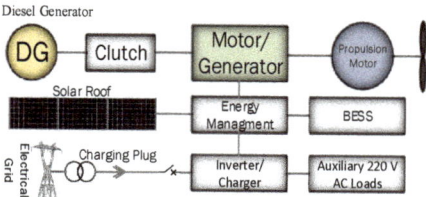

Figure 20. Schematic diagram of hybrid yacht by *Greenline 33*.

The Vision of the Fjords, a diesel-electric hybrid vessel owned and operated by the Fjords. ABB's onboard DC system is installed on the vessel that controls and manages the flow of energy between propeller, diesel engine, and the charging station [124]. The ferry carries around 400 tourists, it cruises between Flåm and Gudvangen, a distance of approximately 32 km. On the route from Flåm to the start of Nærøyfjord, it works on a diesel engine with a speed of around 18–19 knots for a duration of around 30–40 min. From the Nærøyfjord (UNESCO world's heritage-listed place) to Gudvangen, it will switch to a battery resulting in the speed around 8 knots such that people can enjoy the scenes in almost complete silence, the basic schematic diagram of the vessel is depicted in Figure 21. The power system in the vessel comprises of 2 main engines, 2 electric motors, lithium-ion-based ESS, and an onboard 825 V DC. The batteries are connected with a manual plug to the grid and it takes around 20 min to charge it. *Future of the Fjords* [125], a sister vessel of *The Vision of the Fjords* is fully electric vessel, which makes the vessel quiet, vibrationless, and emission-free. For charging the batteries a battery bank is installed in a floating glass at Gudvangen as the capacity of a nearby grid is quite limited. Two lithium-ion-based battery packs for the propulsion motors are installed. It is estimated that the ferry's electricity consumption is around 700 kWh per trip which is approximately equivalent to 80 litres of diesel.

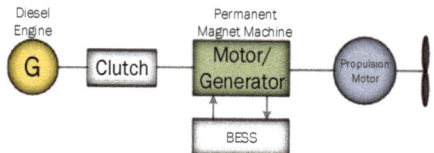

Figure 21. *The Vision of the Fjords* schematic diagram.

M/F Finnøy, a car ferry that was originally a diesel-electric ferry built in 1999 with a capacity of 350 passengers and 110 cars, the ferry runs on the crossing between Oanes and Lauvvik, Norway.

The vessel was upgraded in 2013 with an energy storage system, particularly a lithium polymer-based battery system is installed [126]. Presently, it is owned by Norled, which consists of Siemens based drive system, four diesel-based generators, a battery storage system, and main propulsion system as shown in Figure 22.

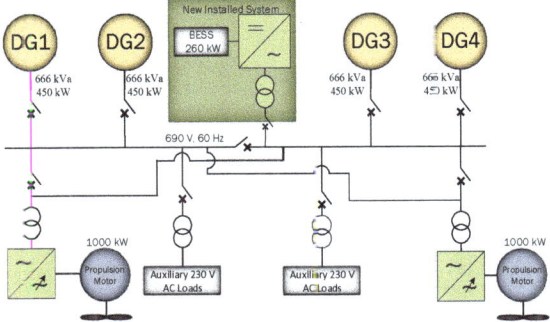

Figure 22. *M/F Finnøy* schematic diagram.

Tycho Brahe and *Aurora*, car ferries operated by HH Ferries group, which sails between Helsingø, Denmark and Helsingborg, Sweden, a distance of around 4 km. *Tycho Brahe* is sailing since 1991 and has the capability of carrying 1250 passengers, 240 cars, 260 trucks, and 9 passenger train coaches at a time. Previously, the ferry used to be powered with traditional power generation sources, i.e., diesel-based generators; 4 diesel engines were installed to cater to daily operations on the vessel. The ferry was docked in June 2017 to re-commission it as an entirely battery-powered vessel. The propulsion system and power for these two ferries are supplied by ABB, which comprises of batteries, an energy storage system, onboard DC grid technology, and control system [127]. The batteries installed on top of the ferry are 640 in number, having the energy density of each battery 6.4 kWh along with 2 deckhouses for transformers, cooling of batteries, and converters. Two generators out of four were removed from the ferry, rest of two generators act as a backup power source, i.e., not be used for daily operation purposes. Furthermore, ABB-based robotic arm is installed, which connects the batteries with the grid every time when the ferry is in port in order to optimize connection time. It is estimated that the emissions of carbon monoxide, carbon dioxide, and sulphur oxide are reduced by 50% [128].

Port-liner, a Dutch company builds an emission-free sailing barges, which are crewless and are operated from the ports of Amsterdam, Rotterdam, and Antwerp [129]. The company further develops an "e-Powerbox", which is vibration and shock free. It can be easily swapped with the charged "e-Powerbox" when the barge is at the port. The electric motors of the barge will be powered by 20-foot batteries. As a consequence of these zero-emission barges, 23,000 trucks that are powered by diesel engines will subsequently be removed from the roads. As, presently in Europe, 74.9% freight is transported by roads, 18.4% by railways, and 6.7% by seaways according to Eurostat statistics. It is expected that five barges will be in operation soon, which have the capability to carry 20 containers, each container's size is around 20 feet. These barges will be fitted with "e-Powerbox" that has the capability to provide power for 15 h. *FCS Alsterwasser*, a first passenger tourist vessel with zero-emissions was developed in Germany that operates on the Alster river. It has a hybrid electric propulsion system with two 50 kW PEM-based fuel cell and a lead-acid based battery system as shown in Figure 23. Table 8 summarizes the type and rating of BESS used in different vessels.

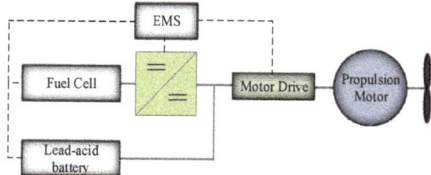

Figure 23. Fuel cell ship (FCS) alsterwasser hybrid battery/fuel cell system [130].

Table 8. Summary of BESS-based vessels.

Vessel's Name	Type of Battery Used	Power Rating	Generation Sources	Type of Ship	Reference
MF Ampere	Lithium-polymer	1040 kW	Battery only	Ferry	[120,121]
Guangzhou (Cargo Ship)	Lithium-ion	2400 kW	Battery only	Cargo Ship	[122]
Greenline 33	Lithium-ion	11.5 kW	Battery + Diesel generator	Yacht	[123]
Greenline 40	Lithium-ion	23 kW	Battery only	Yacht	[123]
Greenline 46	Lithium-ion	46 kW	Battery only	Yacht	[123]
The Vision of the Fjords	Lithium-ion	576 kW	Battery + Diesel generator	Ferry	[124]
Future of the Fjords	Lithium-ion	1800 kW	Battery only	Ferry	[125]
M/F Finnøy	Lithium-polymer	260 kW	Battery + Diesel generator	Ferry	[126]
Tycho Brahe & Aurora	Lithium-ion	4160 kW	Battery only	Ferry	[127]
FCS Alsterwasser	Lead-acid	200 kW	Battery + FC	Ferry	[69]

5. Challenges of Integrating Energy Storage System in Shipboard Microgrids

Electric ships experience immense propulsion load fluctuations on their drive shaft, particularly due to rotational motion of the propeller and waves, which affect the reliability and can cause wear and tear. Hence, modern shipboard microgrids are needed to be designed while considering challenging performance criteria and also considering the environmental concerns. These requirements demand to improve the design methods for vessels and their operation. The ESS can be considered either as the main source of power or as a redundant power source. In literature, there are several works where energy storage has been utilized in terrestrial microgrids to minimize the effects of changes in loads on the crucial parameters of the system. However, in shipboard microgrids, such approaches are yet to be applied at such level. Recently, control techniques being used are adaptive control, particle swarm optimization, proportional integral (PI) control, active and reactive power (PQ) control, etc. The abrupt changes in shiploads due to dynamic pulse loads such as high-power radars, an electromagnetic rail gun, laser self-defense system, etc., changes the power demand in a quick manner. These issues may reduce the efficiency of the whole power system if properly not handled.

Furthermore, ESS technologies are quite expensive and rely on power conversion devices depending on the power system either AC or DC. In this case, a solution might be to install ESS as part of motor drives as shown in Figure 17 in order to eliminate the requirement of additional power conversion devices, hence resulting in the reduction in the cost and weight. In this scenario, by installing ESS alongside with an AFE, the ESS can attain application flexibility, therefore can mitigate harmonics, peak shaving, etc. The battery packs are quite heavy and take a lot of space but can replace at least one prime mover from the vessel. The other issue with battery technology includes the lifetime, swift charging and discharging of batteries may result in heat, which further causes a reduction in the lifetime of a battery. Hence, there might be a possibility that batteries may be defected or died before they manage to cover the installation cost by reducing the consumption of fuel. To solve this problem, the involvement of ultra-capacitors, flywheel, fuel cells, etc., can be beneficial. They are installed with the battery packs to improve charge and discharge speed, increase lifetime, enhance power density and so on. In present, using a single energy storage system might not be a solution as batteries can only provide higher energy density whereas flywheel, ultra-capacitor can provide higher power density. Batteries further have a short life cycle as compared to higher power density-based energy sources. Hence, hybridizing two energy storage devices might be an interesting solution for future shipboard microgrids.

6. Conclusions

This paper reviews different hybrid combinations of energy storage systems for shipboard power systems which are applied in the literatures. The possibility of using energy storage systems in load levelling, peak shaving, power smoothing, and power quality improvement are briefly discussed. It is observed that ESS can be useful to flatten the vessel's load profile, to facilitate starting and stopping of generators, and reduces the number of online prime movers. Therefore, in low loading conditions, the ESS charges and during the high load demands, the ESS provides the stored power and hence discharges. Moreover, it helps the prime movers to run at their optimal fuel-consumption efficient point. ESS in peak shaving applications, further economizes the fuel consumption and therefore results in a reduction of emissions by reducing the number of online generators. Among the batteries, Li-ion is the most used battery for shipboard power applications, specifically for ferries that cruise on shorter routes. Furthermore, it can contribute in reducing the fluctuations caused by the propulsion loads. The hybridization of energy storage devices are expected to provide an extra support in future for larger cargo vessels and for larger routes as well. It is found that battery-flywheel and battery-ultracapacitors energy storage systems have been among the most used energy storage devices, particularly for the applications that are related to shipboard power systems. The hybrid battery-fuel cell is also among the frequently used technologies in literature, but efficiency and storing issues of hydrogen in case of fuel cells is still a major concern.

Author Contributions: The authors have equally contributed to the writing of the manuscript.

Funding: This research received no external funding.

Conflicts of Interest: The authors declare no conflict of interest.

Abbreviations

The following abbreviations are used in this manuscript:

IMO	International marine organization
ESS	Energy storage system
HESS	Hybrid energy storage system
IPS	Integrated power system
VSD	Variable speed drives
AES	All electric ships
SMES	Superconducting magnetic energy storage
FC	Fuel Cell
UPS	Uninterruptible power supply
PEM	Polymer exchange membrane
DP	Dynamic positioning
BESS	Battery energy storage system
FESS	Flywheel energy storage system
UPS	Uninterruptible power supply
MVDC	Medium voltage direct current

References

1. McCoy, T.J. Electric ships past, present, and future [technology leaders]. *IEEE Electr. Mag.* **2015**, *3*, 4–11. [CrossRef]
2. Zahedi, B.; Norum, L.E.; Ludvigsen, K.B. Optimized efficiency of all-electric ships by dc hybrid power systems. *J. Power Sources* **2014**, *255*, 341–354. [CrossRef]
3. Doerry, N. Naval Power Systems: Integrated power systems for the continuity of the electrical power supply. *IEEE Electr. Mag.* **2015**, *3*, 12–21. [CrossRef]
4. Doerry, N.; Amy, J.; Krolick, C. History and the status of electric ship propulsion, integrated power systems, and future trends in the US Navy. *Proc. IEEE* **2015**, *103*, 2243–2251. [CrossRef]
5. International Maritime Organization (IMO). *Third IMO Greenhouse Gas Study*; International Maritime Organization (IMO): London, UK, 2014.

6. International Maritime Organization: Prevention of Air Pollution from Ships. Available online: http://www.imo.org/en/OurWork/Environment/PollutionPrevention/AirPollution/Pages/Air-Pollution.aspx (accessed on 27 August 2018).
7. EU Commission. The Paris Protocol–A Blueprint for Tackling Global Climate Change Beyond 2020. Available online: https://www.eesc.europa.eu/sites/default/files/resources/docs/15_362-ppaper_changement-clim_en.pdf (accessed on 27 August 2018).
8. Johansen, T.A.; Bø, T.I.; Mathiesen, E.; Veksler, A.; Sørensen, A.J. Dynamic positioning system as dynamic energy storage on diesel-electric ships. *IEEE Trans. Power Syst.* **2014**, *29*, 3086–3091. [CrossRef]
9. Veksler, A.; Johansen, T.A.; Skjetne, R. Thrust allocation with power management functionality on dynamically positioned vessels. In Proceedings of the American Control Conference (ACC), Fairmont Queen Elizabeth, Montréal, QC, Canada, 27–29 June 2012; pp. 1468–1475.
10. Li, X.; Hui, D.; Lai, X. Battery energy storage station (BESS)-based smoothing control of photovoltaic (PV) and wind power generation fluctuations. *IEEE Trans. Sustain. Energy.* **2013**, *4*, 464–473. [CrossRef]
11. Hou, J.; Sun, J.; Hofmann, H. Interaction analysis and integrated control of hybrid energy storage and generator control system for electric ship propulsion. In Proceedings of the American Control Conference (ACC), Chicago, IL, USA, 1–3 July 2015; pp. 4988–4993.
12. Haseltalab, A.; Negenborn, R.R.; Lodewijks, G. Multi-level predictive control for energy management of hybrid ships in the presence of uncertainty and environmental disturbances. *IFAC-PapersOnLine* **2016**, *49*, 90–95. [CrossRef]
13. Hebner, R.E.; Davey, K.; Herbst, J.; Hall, D.; Hahne, J.; Surls, D.D.; Ouroua, A. Dynamic load and storage integration. *Proc. IEEE* **2015**, *103*, 2344–2354. [CrossRef]
14. Hemmati, R.; Saboori, H. Emergence of hybrid energy storage systems in renewable energy and transport applications–A review. *Renew. Sustain. Energy Rev.* **2016**, *65*, 11–23. [CrossRef]
15. Chen, H.; Cong, T.N.; Yang, W.; Tan, C.; Li, Y.; Ding, Y. Progress in electrical energy storage system: A critical review. *Prog. Nat. Sci.* **2009**, *19*, 291–312. [CrossRef]
16. Argyrou, M.C.; Christodoulides, P.; Kalogirou, S.A. Energy storage for electricity generation and related processes: Technologies appraisal and grid scale applications. *Renew. Sustain. Energy Rev.* **2018**, *94*, 804–821. [CrossRef]
17. Stan, A.T.; Swierczynski, M.; Stroe, D.I.; Teodorescu, R.; Andreasen, S.J.; Moth, K. A comparative study of lithium ion to lead acid batteries for use in UPS applications. In Proceedings of the Telecommunications Energy Conference (INTELEC), Vancouver, BC, Canada, 28 September–2 October 2014; pp. 1–8.
18. Nilsson, A.O. Nickel cadmium batteries in UPS design features. In Proceedings of the Telecommunications Energy Conference, San Diego, CA, USA, 30 October–2 November 1988; pp. 388–393.
19. Harrison, A.I. Batteries and AC phenomena in UPS systems: the battery point of view. In Proceedings of the Telecommunications Energy Conference, Florence, Italy, 15–18 October 1989; pp. 5–12.
20. Bekiarov, S.B.; Nasiri, A.; Emadi, A. A new reduced parts on-line single-phase UPS system. In Proceedings of the Industrial Electronics Society, IECON'03, Roanoke, VA, USA, 2–6 November 2003; Volume 1, pp. 688–693.
21. Zaghib, K.; Charest, P.; Guerfi, A.; Shim, J.; Perrier, M.; Striebel, K. Safe Li-ion polymer batteries for HEV applications. *J. Power Sources* **2004**, *134*, 124–129. [CrossRef]
22. Karden, E.; Ploumen, S.; Fricke, B.; Miller, T.; Snyder, K. Energy storage devices for future hybrid electric vehicles. *J. Power Sources* **2007**, *168*, 2–11. [CrossRef]
23. Burke, A.F. Batteries and ultracapacitors for electric, hybrid, and fuel cell vehicles. *Proc. IEEE* **2007**, *95*, 806–820. [CrossRef]
24. Zhao, H.; Wu, Q.; Hu, S.; Xu, H.; Rasmussen, C.N. Review of energy storage system for wind power integration support. *Appl. Energy* **2015**, *137*, 545–553. [CrossRef]
25. Chatzivasileiadi, A.; Ampatzi, E.; Knight, I. Characteristics of electrical energy storage technologies and their applications in buildings. *Renew. Sustain. Energy Rev.* **2013**, *25*, 814–830. [CrossRef]
26. Ibrahim, H.; Ilinca, A.; Perron, J. Energy storage systems characteristics and comparisons. *Renew. Sustain. Energy Rev.* **2008**, *12*, 1221–1250. [CrossRef]
27. Vazquez, S.; Lukic, S.M.; Galvan, E.; Franquelo, L.G.; Carrasco, J.M. Energy storage systems for transport and grid applications. *IEEE Trans. Ind. Electr.* **2010**, *57*, 3881–3895. [CrossRef]

28. Dufo-López, R.; Lujano-Rojas, J.M.; Bernal-Agustín, J.L. Comparison of different lead-acid battery lifetime prediction models for use in simulation of stand-alone photovoltaic systems. *Appl. Energy* **2014**, *115*, 242–253. [CrossRef]
29. Parker, C.D. Lead-acid battery energy-storage systems for electricity supply networks. *J. Power Sources* **2001**, *100*, 18–28. [CrossRef]
30. Mohod, S.W.; Aware, M.V. Micro wind power generator with battery energy storage for critical load. *IEEE Syst. J.* **2012**, *6*, 18–125. [CrossRef]
31. Lam, L.T. Louey, R. Development of ultra-battery for hybrid-electric vehicle applications. *J. Power Sources* **2006**, *158*, 1140–1148. [CrossRef]
32. Olson, J.B.; Sexton, E.D. Operation of lead-acid batteries for HEV applications. In Proceedings of the Battery Conference on Applications and Advance, Long Beach, CA, USA, 11–14 January 2000; pp. 205–210.
33. Hadjipaschalis, I.; Poullikkas, A.; Efthimiou, V. Overview of current and future energy storage technologies for electric power applications. *Renew. Sustain. Energy Rev.* **2009**, *13*, 1513–1522. [CrossRef]
34. Lukic, S.M.; Cao, J.; Bansal, R.C.; Rodriguez, F.; Emadi, A. Energy storage systems for automotive applications. *IEEE Trans. Ind. Electron.* **2008**, *55*, 2258–2267. [CrossRef]
35. Wehrey, M.C. What's new with hybrid electric vehicles. *IEEE Power Energy Mag.* **2004**, *2*, 34–39. [CrossRef]
36. Venkatasetty, H.V.; Jeong, Y.U. Recent advances in lithium-ion and lithium-polymer batteries. In Proceedings of the Battery Conference on Applications and Advances, Long Beach, CA, USA, 18 January 2002; pp. 173–178.
37. Khaligh, A.; Li, Z. Battery, ultracapacitor, fuel cell, and hybrid energy storage systems for electric, hybrid electric, fuel cell, and plug-in hybrid electric vehicles: State of the art. *IEEE Trans. Veh. Technol.* **2010**, *59*, 2806–2814. [CrossRef]
38. Cao, J.; Emadi, A. A new battery/ultracapacitor hybrid energy storage system for electric, hybrid, and plug-in hybrid electric vehicles. *IEEE Trans. Power Electron.* **2012**, *27*, 122–132.
39. Anstrom, J.R.; Zile, B.; Smith, K.; Hofmann, H.; Batra, A. Simulation and field-testing of hybrid ultra-capacitor/battery energy storage systems for electric and hybrid-electric transit vehicles. In Proceedings of the Applied Power Electronics Conference and Exposition (APEC), Charlotte, NC, USA, 15–19 March 2015; Volume 1, pp. 491–497.
40. Nájera, J.; Moreno-Torres, P.; Lafoz, M.; de Castro, R.M.; Arribas, J.R. Approach to hybrid energy storage systems dimensioning for urban electric buses regarding efficiency and battery aging. *Energies* **2017**, *10*, 1708. [CrossRef]
41. Kouchachvili, L.; Yaïci, W.; Entchev, E. Hybrid battery/supercapacitor energy storage system for the electric vehicles. *J. Power Sources* **2018**, *374*, 237–248. [CrossRef]
42. Ultracapacitor Based Buses. Available online: https://www.treehugger.com/cars/ultracapacitor-buses-work-as-long-as-you-have-lots-of-quick-charge-stations.html (accessed on 27 August 2018).
43. Park, J.D.; Kalev, C.; Hofmann, H.F. Control of high-speed solid-rotor synchronous reluctance motor/generator for flywheel-based uninterruptible power supplies. *IEEE Trans. Ind. Electron.* **2008**, *55*, 3038–3046. [CrossRef]
44. Cheng, M.; Sami, S.S.; Wu, J. Benefits of using virtual energy storage system for power system frequency response. *Appl. Energy* **2017**, *194*, 376–385. [CrossRef]
45. Díaz-González, F.; Sumper, A.; Gomis-Bellmunt, O.; Bianchi, F.D. Energy management of flywheel-based energy storage device for wind power smoothing. *Appl. Energy* **2013**, *110*, 207–219. [CrossRef]
46. Spiryagin, M.; Wolfs, P.; Szanto, F.; Sun, Y.Q.; Cole, C.; Nielsen, D. Application of flywheel energy storage for heavy haul locomotives. *Appl. Energy* **2015**, *157*, 607–618. [CrossRef]
47. McGroarty, J.; Schmeller, J.; Hockney, R.; Polimeno, M. Flywheel energy storage system for electric start and an all-electric ship. In Proceedings of the Electric Ship Technologies Symposium, Philadelphia, PA, USA, 25–27 July 2005; pp. 400–406.
48. Anvari-Moghaddam, A.; Dragicevic, T.; Meng, L.; Sun, B.; Guerrero, J.M. Optimal planning and operation management of a ship electrical power system with energy storage system. In Proceedings of the IECON 42nd Annual Conference of the IEEE Industrial Electronics Society, Florence, Italy, 24–27 October 2016; pp. 2095–2099.
49. Domaschk, L.N.; Ouroua, A.; Hebner, R.E.; Bowlin, O.E.; Colson, W.B. Coordination of large pulsed loads on future electric ships. *IEEE Trans. Magn.* **2007**, *43*, 450–455. [CrossRef]

50. Hebner, R.E.; Herbst, J.D.; Gattozzi, A.L. Pulsed power loads support and efficiency improvement on navy ships. *Naval Eng. J.* **2010**, *122*, 23–32. [CrossRef]
51. Jeong, H.W.; Kim, Y.S.; Kim, C.H.; Choi, S.H.; Yoon, K.K. Analysis on application of flywheel energy storage system for offshore plants with dynamic positioning system. *J. Korean Soc. Mar. Eng.* **2012**, *36*, 935–941. [CrossRef]
52. Jeong, H.W.; Ha, Y.S.; Kim, Y.S.; Kim, C.H.; Yoon, K.K.; Seo, D.H. Shore power to ships and offshore plants with flywheel energy storage system. *J. Korean Soc. Mar. Eng.* **2013**, *37*, 771–777. [CrossRef]
53. Boom, R.; Peterson, H. Superconductive energy storage for power systems. *IEEE Trans. Magn.* **1972**, *8*, 701–703. [CrossRef]
54. Hsony, W.M.; Dodds, S.J. Applied superconductivity developments in Japan. *Power Eng. J.* **1993**, *7*, 170–176. [CrossRef]
55. Holla, R.V. Energy Storage Methods-Superconducting Magnetic Energy Storage—A Review. *J. Undergrad. Res. Univ. Ill. Chic.* **2015**, *8*. [CrossRef]
56. Beach, F.C.; McNab, I.R. Present and future naval applications for pulsed power. In Proceedings of the Pulsed Power Conference, Monterey, CA, USA, 13–15 June 2005; pp. 1–7.
57. Wolfe, T.; Riedy, P.; Drake, J.; MacDougall, F.; Bernardes, J. Preliminary design of a 200 MJ pulsed power system for a naval railgun proof of concept facility. In Proceedings of the Pulsed Power Conference, Monterey, CA, USA, 13–15 June 2005; pp. 70–74.
58. Miller, J.; Santosusso, D.; Uva, M.; Woods, K.; Fitzpatrick, B. Naval superconducting integrated power system (sips). In Proceedings of the 10th Intelligent Ship Symposium, Philadelphia, Pennsylvania, 22–23 May 2013.
59. Superconducting Magnetic Energy Storage. Available online: https://apps.dtic.mil/dtic/tr/fulltext/u2/a338581.pdf (accessed on 27 August 2018).
60. German HDW Submarine. Available online: https://www.naval-technology.com/projects/type_212/ (accessed on 27 August 2018).
61. Fuel Cell Ship Alsterwasser. Available online: https://www.drewsmarine.com/en/references/passenger-ferries/fcs-alsterwasser (accessed on 27 August 2018).
62. Symington, W.P.; Belle, A.; Nguyen, H.D.; Binns, J.R. Emerging technologies in marine electric propulsion. *Proc. Inst. Mech. Eng. Part M J. Eng. Marit. Environ.* **2016**, *230*, 187–198. [CrossRef]
63. Carlton, J.; Aldwinkle, J.; Anderson, J. *Future Ship Powering Options: Exploring Alternative Methods of Ship Propulsion*; Royal Academy of Engineering: London, UK, 2013.
64. Comparison of Fuel Cell Technologies. Available online: https://www1.eere.energy.gov/hydrogenandfuelcells/fuelcells/pdfs/fc_comparison_chart.pdf (accessed on 27 August 2018).
65. Viking Lady. Available online: https://www.wartsila.com/resources/customer-references/view/viking-lady (accessed on 27 August 2018).
66. Lloyd, G. DNV, Shipping Industry Eyeing Hydrogen Fuel Cells as Possible Pathway to Emissions Reduction. Available online: http://www.greencarcongress.com/2012/09/h2shipping-20120907.html (accessed on 27 August 2018).
67. Viking Lady. Available online: http://maritimeinteriorpoland.com/references/viking-lady/ (accessed on 27 August 2018).
68. Fuel Cell Boat (Nemo H2). Available online: http://www.opr-advies.nl/fuelcellboat/efcbboot.html (accessed on 27 August 2018).
69. Henderson, K. Fuel Cell Vessel Back in Service. Available online: http://articles.maritimepropulsion.com/article/Fuel-Cell-Vessel-Back-In-Service80232.aspx (accessed on 27 August 2018).
70. SF-BREEZE. Available online: https://energy.sandia.gov/transportation-energy/hydrogen/market-transformation/maritime-fuel-cells/sf-breeze/ (accessed on 27 August 2018).
71. Pa-x-ell. Available online: http://www.e4ships.de/aims-35.html (accessed on 27 August 2018).
72. METHAPU Prototypes Methanol SOFC for Ships. *Fuel Cells Bull.* **2008**, *5*, 4–5. 2859(08)70190-1. [CrossRef]
73. SFC Fuel Cells for US Army, Major Order from German Military. *Fuel Cells Bull.* **2012**, *6*, 4. [CrossRef]
74. Jafarzadeh, S.; Schjølberg, I. Emission Reduction in Shipping Using Hydrogen and Fuel Cells. In Proceedings of the ASME International Conference on Ocean, Offshore and Arctic Engineering, Trondheim, Norway, 25–30 June 2017; p. V010T09A011.
75. MS Forester. Available online: https://shipandbunker.com/news/emea/914341-fuel-cell-technology-successfully-tested-on-two-vessels (accessed on 27 August 2018).

76. 212A Class Submarine. Available online: http://www.seaforces.org/marint/German-Navy/Submarine/Type-212A-class.htm (accessed on 27 August 2018).
77. SSK S-80 Class Submarine. Available online: https://www.naval-technology.com/projects/ssk-s-80-class-submarine/ (accessed on 27 August 2018).
78. Kumm, W.H.; Lisie, E.L., Jr. *Feasibility Study of Repowering the USCGC VINDICATOR (WMEC-3) with Modular Diesel Fueled Direct Fuel Cells*; Arctic Energies Ltd Severna Park MD: Groton, MA, USA, 1997.
79. Tang, Y.; Khaligh, A. On the feasibility of hybrid battery/ultracapacitor energy storage systems for next generation shipboard power systems. In Proceedings of the Vehicle Power and Propulsion Conference (VPPC), Lille, France, 1–3 September 2010; pp. 1–6.
80. Hou, J.; Sun, J.; Hofmann, H.F. Mitigating power fluctuations in electric ship propulsion with hybrid energy storage system: Design and analysis. *IEEE J. Ocean. Eng.* **2018**, *43*, 93–107. [CrossRef]
81. Alafnan, H.; Zhang, M.; Yuan, W.; Zhu, J.; Li, J.; Elsheikh, M.; Li, X. Stability Improvement of DC Power Systems in an All-Electric Ship Using Hybrid SMES/Battery, *IEEE Trans. Appl. Supercond.* **2018**, *28*, 1–6. [CrossRef]
82. Hou, J.; Sun, J.; Hofmann, H. Control development and performance evaluation for battery/flywheel hybrid energy storage solutions to mitigate load fluctuations in all-electric ship propulsion systems. *Appl. Energy* **2018**, *212*, 919–930. [CrossRef]
83. Hou, J.; Sun, J.; Hofmann, H. Battery/flywheel Hybrid Energy Storage to mitigate load fluctuations in electric ship propulsion systems. In Proceedings of the American Control Conference (ACC), Seattle, WA, USA, 24–26 May 2017; pp. 1296–1301.
84. Sander, M.; Gehring, R.; Neumann, H.; Jordan, T. LIQHYSMES storage unit-Hybrid energy storage concept combining liquefied hydrogen with Superconducting Magnetic Energy Storage. *Int. J. Hydrog. Energy* **2012**, *37*, 14300–14306. [CrossRef]
85. Kisacikoglu, M.C.; Uzunoglu, M.; Alam, M.S. Load sharing using fuzzy logic control in a fuel cell/ultracapacitor hybrid vehicle. *Int. J. Hydrog. Energy* **2009**, *34*, 1497–1507. [CrossRef]
86. Roda, V.; Carroquino, J.; Valiño, L.; Lozano, A.; Barreras, F. Remodeling of a commercial plug-in battery electric vehicle to a hybrid configuration with a PEM fuel cell. *Int. J. Hydrog. Energy* **2018**, *43*, 16959–16970. [CrossRef]
87. Zhao, P.; Dai, Y.; Wang, J. Design and thermodynamic analysis of a hybrid energy storage system based on A-CAES (adiabatic compressed air energy storage) and FESS (flywheel energy storage system) for wind power application. *Energy* **2014**, *70*, 674–684. [CrossRef]
88. Lemofouet, S.; Rufer, A. Hybrid energy storage system based on compressed air and super-capacitors with maximum efficiency point tracking (MEPT). *IEEE Trans. Ind. Appl.* **2006**, *126*, 911–920. [CrossRef]
89. Cericola, D.; Kötz, R. Hybridization of rechargeable batteris and electrochemical capacitors principles and limits, *Electrochim. Acta* **2012**, *72*, 1–17.
90. Cooper, A.; Furakawa, J.; Lam, L.; Kellaway, M. The UltraBattery—A new battery design for a new beginning in hybrid electric vehicle energy storage. *J. Power Sources* **2009**, *188*, 642–649. [CrossRef]
91. Liu, H.; Zhang, Q.; Qi, X.; Han, Y.; Lu, F. Estimation of PV output power in moving and rocking hybrid energy marine ships. *Appl. Energy* **2017**, *204*, 362–372. [CrossRef]
92. Balsamo, F.; Capasso, C.; Miccione, G.; Veneri, O. Hybrid Storage System Control Strategy for All-Electric Powered Ships. *Energy Procedia* **2017**, *126*, 1083–1090. [CrossRef]
93. Cohen, I.J.; Westenhover, C.S.; Wetz, D.A.; Heinzel, J.M.; Dong, Q. Evaluation of an actively controlled battery-capacitor hybrid energy storage module (HESM) for use in driving pulsed power applications. In Proceedings of the Pulsed Power Conference (PPC), Austin, TX, USA, 31 May–4 June 2015; pp. 1–5.
94. Vu, T.V.; Gonsoulin, D.; Diaz, F.; Edrington, C.S.; El-Mezyani, T. Predictive Control for Energy Management in Ship Power Systems Under High-Power Ramp Rate Loads. *IEEE Trans. Energy Convers.* **2017**, *32*, 783–797. [CrossRef]
95. Kulkarni, S.; Santoso, S. Impact of pulse loads on electric ship power system: With and without flywheel energy storage systems. In Proceedings of the Electric Ship Technologies Symposium, Baltimore, MD, USA, 20–22 April 2009; pp. 568–573.
96. Elsayed, A.T.; Mohammed, O.A. A comparative study on the optimal combination of hybrid energy storage system for ship power systems. In Proceedings of the Electric Ship Technologies Symposium (ESTS), Alexandria, VA, USA, 21–24 June 2015; pp. 140–144.

97. Lan, H.; Bai, Y.; Wen, S.; Yu, D.C.; Hong, Y.Y.; Dai, J.; Cheng, P. Modeling and stability analysis of hybrid pv/diesel/ess in ship power system. *Inventions* **2016**, *1*, 5. [CrossRef]
98. Mouritz, A.P; Gellert, E.; Burchill, P.; Challis, K. Review of advanced composite structures for naval ships and submarines. *Composite structures* **2001**, *53*, 21–42. [CrossRef]
99. Han, J.; Charpentier, J.F.; Tang, T. State of the Art of Fuel Cells for Ship applications. In Proceedings of the IEEE International Symposium on Industrial Electronics (ISIE), Hangzhou, China, 28–31 May 2012; pp. 1456–1461.
100. Psoma, A.; Sattler, G. Fuel cell systems for submarines: from the first idea to serial production. *J. Power Sources* **2002**, *106*, 381–383. [CrossRef]
101. Sattler, G. PEFCs for naval ships and submarines: Many tasks, one solution. *J. Power Sources* **1998**, *71*, 144–149. [CrossRef]
102. Luckose, L.; Urlaub, N.J.; Wiedeback, N.J.; Hess, H.L.; Johnson, B.K. Proton exchange membrane fuel cell (pemfc) modeling in pscad/emtdc. In Proceedings of the Electrical Power and Energy Conference (EPEC), Winnipeg, MB, Canada, 3–5 October 2011; pp. 11–16.
103. Abkenar, A.T.; Nazari, A.; Jayasinghe, S.D.G.; Kapoor, A.; Negnevitsky, M. Fuel cell power management using genetic expression programming in all-electric ships. *IEEE Trans. Energy Convers.* **2017**, *32*, 779–787. [CrossRef]
104. Su, C.L.; Weng, X.T.; Chen, C.J. Power generation controls of fuel cell/energy storage hybrid ship power systems. In Proceedings of the Transportation Electrification Asia-Pacific (ITEC Asia-Pacific), Beijing, China, 31 August–3 September 2014; pp. 1–6.
105. Eyer, J.M.; Erdman, B.; Iannucci, J.J., Jr. *Innovative Applications of Energy Storage in a Restructured Electricity Marketplace: Phase III Final Report: A Study for the DOE Energy Storage Systems Program (No. SAND2003-2546)*; Sandia National Laboratorie: Albuquerque, NM, USA, 2005.
106. Energy Storage System for Ships-Work in Progress. Available online: http://corvusenergy.com/work-in-progress/ (accessed on 20 September 2018).
107. Mattern, K.; Ellis, A.; Williams, S.E.; Edwards, C.; Nourai, A.; Porter, D. Application of inverter-based systems for peak shaving and reactive power management. In Proceedings of the Transmission and Distribution Conference and Exposition, Chicago, IL, USA, 21–24 April 2008; pp. 1–4.
108. Sonoda, N.; Matsunaga, H.; Gengo, T.; Minami, M.; Oishi, M.; Hashimoto, T. Development of Containerized Energy Storage System with Lithium-ion batteries. *Mitsubishi Heavy Ind. Tech. Rev.* **2013**, *50*, 36–41.
109. Bø, T.I.; Johansen, T.A. Battery power smoothing control in a marine electric power plant using nonlinear model predictive control. *IEEE Trans. Control Syst. Technol.* **2017**, *25*, 1449–1456. [CrossRef]
110. Det Norske Veritas. Rules for Classification and Construction. Ship Technology, Seagoing Ships, Electrical Installations of Ships. Available online: https://rules.dnvgl.com/docs/pdf/DNV/rulesship/2012-01/ts408.pdf (accessed on 27 August 2018).
111. Tang, J.; Xiong, B.; Huang, Y.; Yuan, C.; Su, G. Optimal configuration of energy storage system based on frequency hierarchical control in ship power system with solar photovoltaic plant. *J. Eng.* **2017**, *13*, 1511–1514. [CrossRef]
112. Shagar, V.; Jayasinghe, S.G.; Enshaei, H. Frequency transient suppression in hybrid electric ship power systems: A model predictive control strategy for converter control with energy storage. *Inventions* **2018**, *3*, 13. [CrossRef]
113. Kim, S.Y.; Choe, S.; Ko, S.; Sul, S.K. A Naval Integrated Power System with a Battery Energy Storage System: Fuel efficiency, reliability, and quality of power. *IEEE Electr. Mag.* **2015**, *3*, 22–33. [CrossRef]
114. Terriche, Y.; Kerdoun, D.; Djeghloud, H. A new passive compensation technique to economically improve the power quality of two identical single-phase feeders. In Proceedings of the IEEE 15th International Conference on Environment and Electrical Engineering (EEEIC), Rome, Italy, 10–13 June 2015; pp. 54–59.
115. Terriche, Y.; Golestan, S.; Guerrero, J.M.; Kerdoune, D.; Vasquez, J.C. Matrix pencil method-based reference current generation for shunt active power filters. *IET Power Electron.* **2017**, *11*, 772–780. [CrossRef]
116. Tarasiuk, T. Comparative study of various methods of DFT calculation in the wake of IEC Standard 61000-4-7. *IEEE Trans. Instrum. Meas.* **2009**, *58*, 3666–3677. [CrossRef]
117. Xie, C.; Zhang, C. Research on the ship electric propulsion system network power quality with flywheel energy storage. In Proceedings of the Power and Energy Engineering Conference (APPEEC) Asia-Pacific, Chengdu, China, 28–31 March 2010; pp. 1–3.

118. Samineni, S.; Johnson, B.K.; Hess, H.L.; Law, J.D. Modeling and analysis of a flywheel energy storage system for voltage sag correction. *IEEE Trans. Ind. Appl.* **2006**, *42*, 42–52. [CrossRef]
119. Mo, R.; Li, H. Hybrid energy storage system with active filter function for shipboard MVDC system applications based on isolated modular multilevel DC/DC converter. *IEEE J. Emerg. Sel. Top. Power Electron.* **2017**, *5*, 79–87. [CrossRef]
120. Ampere Ferry, World's First All-Ectric Car Ferry. Available online: https://corvusenergy.com/marine-project/mf-ampere-ferry/ (accessed on 27 August 2018).
121. Corvus Energy. Available online: https://corvusenergy.com/tag/corvus-at6500/ (accessed on 27 August 2018).
122. World's First 2,000-Ton Electric Boat Launched. Available online: http://www.chinadaily.com.cn/china/2017-11/13/content_34470726.htm (accessed on 27 August 2018).
123. Greenline Yachts. Available online: https://www.greenlinehybrid.si (accessed on 27 August 2018).
124. Vision of the Fjords. Avalaible online: https://new.abb.com/marine/references/vision-of-the-fjords (accessed on 27 August 2018).
125. Future of the Fjords. Available online: https://www.braa.no/news/future-of-the-fjords (accessed on 27 August 2018).
126. Finnøy, M.F. Available online: https://corvusenergy.com/portfolio/mf-finnoy/ (accessed on 27 August 2018).
127. HH Ferries—Zero Emission Operation. Available online: https://new.abb.com/marine/references/hh-ferries (accessed on 27 August 2018).
128. Brahe, T. Available online: http://sailwiththecurrent.com/ (accessed on 27 August 2018).
129. The Guardian: World's First Electric Container Barges to Sail from European Ports This Summer. Available online: https://www.portliner.nl/media/news/272284_the-guardian-world-s-first-electric-container-barges-to-sail-from-european-ports-this-summer (accessed on 27 August 2018).
130. Bassam, A.M.; Phillips, A.B.; Turnock, S.R.; Wilson, P.A. Development of a multi-scheme energy management strategy for a hybrid fuel cell driven passenger ship. *Int. J. Hydrog. Energy* **2017**, *42*, 623–635. [CrossRef]

© 2018 by the authors. Licensee MDPI, Basel, Switzerland. This article is an open access article distributed under the terms and conditions of the Creative Commons Attribution (CC BY) license (http://creativecommons.org/licenses/by/4.0/).

Article

A Distributed Control Strategy for Islanded Single-Phase Microgrids with Hybrid Energy Storage Systems Based on Power Line Signaling

Pablo Quintana-Barcia [1,*], **Tomislav Dragicevic** [2], **Jorge Garcia** [1], **Javier Ribas** [1] **and Josep M. Guerrero** [2]

[1] ce3i2 Group, Department of Electrical Engineering, University of Oviedo, 33204 Gijon, Spain; garciajorge@uniovi.es (J.G.); ribas@uniovi.es (J.R.)
[2] Department of Energy Technology, Aalborg University, 9920 Aalborg, Denmark; tdr@et.aau.dk (T.D.); joz@et.aau.dk (J.M.G.)
* Correspondence: quintanapablo@uniovi.es; Tel.: +34-985-18-2557

Received: 11 December 2018; Accepted: 22 December 2018; Published: 28 December 2018

Abstract: Energy management control is essential to microgrids (MGs), especially to single-phase ones. To handle the variety of distributed generators (DGs) that can be found in a MG, e.g., renewable energy sources (RESs) and energy storage systems (ESSs), a coordinated power regulation is required. The latter are generally battery-based systems whose lifetime is directly related to charge/discharge processes, whereas the most common RESs in a MG are photovoltaic (PV) units. Hybrid energy storage systems (HESS) extend batteries life expectancy, thanks to the effect of supercapacitors, but they also require more complex control strategies. Conventional droop methodologies are usually applied to provide autonomous and coordinated power control. This paper proposes a method for coordination of a single-phase MG composed by a number of sources (HESS, RES, etc.) using power line signaling (PLS). In this distributed control strategy, a signal whose frequency is higher than the grid is broadcasted to communicate with all DGs when the state of charge (*SoC*) of the batteries reaches a maximum value. This technique prevents batteries from overcharging and maximizes the power contribution of the RESs to the MG. Moreover, different commands apart from the *SoC* can be broadcasted, just by changing to other frequency bands. The HESS master unit operates as a grid-forming unit, whereas RESs act as grid followers. Supercapacitors in the HESS compensate for energy peaks, while batteries respond smoothly to changes in the load, also expanding its lifetime due to less aggressive power references. In this paper, a control structure that allows the implementation of this strategy in single-phase MGs is presented, with the analysis of the optimal range of PLS frequencies and the required self-adaptive proportional-resonant controllers.

Keywords: active power control; energy storage; hybrid; microgrid; photovoltaic; power-line signaling; renewable energy sources; single-phase

1. Introduction

Power systems of today and those developed more than a century ago have several points in common. Firstly, they consist of large power plants installed far away from consumption points. Power flows are unidirectional, moving through long, expensive transmission lines and their operation is demand-driven. These power systems are exceptionally complex and require reliable control strategies to ensure the quality of the grid [1–3]. In the last few decades, this concept has been continuously evolving thanks to modern solutions such as distributed generators DGs—primarily based on energy storage systems (ESSs) and renewable energy sources (RESs)—active demand management combined with smart control, and the introduction of new communication technologies (ITCs) [4,5]. Researchers have been seeking a robust and trustworthy solution that integrates ESSs, RESs and loads into small power systems. This is what has pushed the emergence of the microgrid (MG) concept [1,3–6].

A popular solution today is s DC MG, due to the fact that there is no need for synchronization and the non-existence of reactive power. However, AC MGs are still a valid and reliable solution [3]. They can be operated either in grid-connected (exchanging power with the mains) or islanded (supporting local loads if the grid is not present) modes, although these changes must be seamless and swift, avoiding undesirable transients [7].

In recent years, photovoltaic (PV) unit installation costs have decreased dramatically, and this technology has become one of the major DGs meant to supply MGs. Small wind turbines are also beginning to carve a niche in the market, albeit more slowly [7–10]. However, due to the stochastic nature of renewable energies, ESSs are essential elements for balancing power flows between RESs and loads in islanded MGs [3,11,12]. Moreover, a master ESS works as a grid-forming unit, generating the same AC grid conditions as conventional power systems, whereas renewable sources usually operate as grid-following systems, injecting all their available power into the MG [5,13]. Conventional ESS also has the role of power balance and frequency stability by absorbing or injecting a current from its power source, i.e., batteries. This concept implies that the capacity limitation of these electrochemical devices must be considered in the studies, and to preserve their lifetime, avoiding frequent deep discharge cycles is crucial. The state of charge (SoC) of the batteries needs to be kept, therefore, in a safe region in order not to damage the devices [14,15]. This is why the hybrid energy storage system (HESS) is becoming an interesting solution, able to extend batteries' useful life. By combining fast-dynamics high-power storage devices as supercapacitors and ultracapacitors with bulk-energy electrochemical units, the performance of classic grid forming ESSs has been improved [16–19]. The initial investment in supercapacitors can be paid off by extending the useful life of the batteries. In this work, the master HESS consists of a battery energy storage system (BESS) plus a supercapacitor energy storage system (SESS).

On the other hand, classic MGs are based on three-phase systems. The advantages of three-phase systems are well known: power delivered is constant, transmission of power requires less conductor material, they exhibit good stability and reliability, etc. Single-phase AC needs more capacitance in the DC link than three-phase, typically electrolytic capacitors that used to reduce lifetime. However, electrolytic capacitors' reliability has increased over the past years and now they are not as critical as they used to be [20]. The key to making them last longer is to have them working under their maximum operating temperature [21–24]. On the other hand, it is a fact that most buildings are single-phase supplied. This implies that a small community of neighbors with a certain number of renewable elements (PV panels) and batteries can become a single-phase MG just by adding some sort of control: the most popular and well-known kind of control of a MG is a centralized structure. All functionalities can be integrated into a MG's central controller, which makes decisions based on the measurements from the sensors all over the power system. After processing the data, the central controller sends instructions to the elements that form the MG through some kind of communications system. e.g., wireless, droop algorithms, wired connection, etc. This offers good control capability, but if the number of units increases, their connectivity may require extensive hardware. In addition, the reliability of the whole system depends on one key element [25,26]. Droop control strategy (using frequency deviation of each unit to distribute active power) is widely accepted to fit into this requirement. However, the active power distribution is based on a unified local control algorithm, which ignores the inherent power regulation difference between the ESS and the RESs [27].

In order to tackle this issue and avoid using external communications, a power line can be employed. This technique provides a distributed control using the MG's own power lines as an interface. Signals travel along these carriers with a certain frequency, providing significant information to all the units that form the MG. However, this implies an introduction of noise and therefore, the bandwidth of these signals must be properly designed [28–32]. Previous works have employed similar techniques for islanded DC or three-phase AC microgrids, control of parallel inverters, and more. However, in the particular case of this work, the BESS generates a power line signal (PLS) that informs the RESs distributed along the single-phase MG to reduce their power contribution, due to the fact that the batteries are reaching their maximum *SoC*. The frequency of this signal is proportional to the *SoC* of the batteries. However, below a certain *SoC*, the PLS is turned off and all the grid-followers operate at their nominal operation point. Additional PLS triggers may be programmed, e.g., protection against huge derivatives of batteries' input current, due to extreme sudden changes in the load, a reactive power command, etc. This flexible solution avoids using centralized control or droop strategies and hence, there is no need for secondary control of the frequency.

This paper is organized as follows: in Section 2, the physical configuration of the MG is presented. Section 3 describes the PLS concept and how it can be applied to single-phase MGs. The energy management strategy of the whole MG and how the renewable sources have to react when the PLS is detected is explained in Section 4. The proposed control strategy is verified in Section 5 through hardware-in-the-loop results. This work is concluded in Section 6, where the obtained results are discussed and conclusions are reached.

2. Single-Phase MG Structure and System Configuration

Figure 1 shows a possible single-phase MG connected to the mains through an intelligent transfer switch (ITS). When a fault occurs on the utility grid, the ITS disconnects the MG to enable islanded operation. Then, RES and HESS units are left on their own to supply the loads at nominal voltage and frequency. This MG is formed around a common AC link, to which the HESS, RESs, and loads are directly connected. PV panels are depicted attached to a maximum power point tracker (MPPT) converter, although they could be directly connected to the AC line through the power inverter. Commonly, loads can be practically divided into active and passive ones, but all of them are usually designed for a wide range of input AC voltage, e.g., 100–240 V RMS.

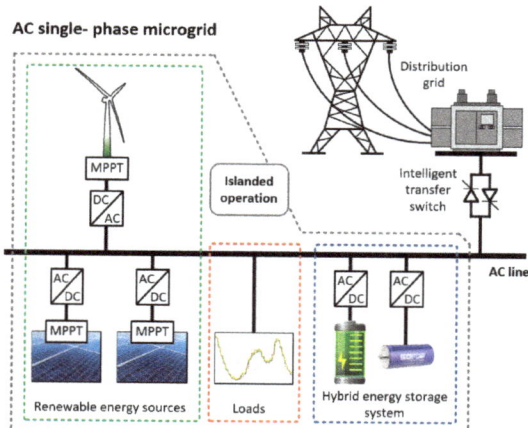

Figure 1. Typical configuration of an AC single-phase microgrid.

As previously mentioned, the energy storage unit fixes both voltage and frequency in the MG during islanded operation, and provides a power buffer, as expressed in (1).

When the *SoC* of the batteries is in a safe region, they can absorb the extra power generated by the RESs that is not consumed by the loads (if there is any), charging up these electrochemical devices.

$$\left. \begin{array}{c} P_{HESS} = P_{BESS} + P_{SESS} \\ P_{HESS} = \sum_{j=1}^{m} P_{LOAD_j} - \sum_{i=1}^{n} P_{RES_i} \end{array} \right\} \quad (1)$$

Under these conditions, the RESs operate by injecting all the available power with an MPPT algorithm. Different up to date control strategies have been developed for both PV and wind sources [9,10,33–35]. For this work, the authors have focused their interests on PV technology and the perturb and observe method. Like all MMPT, this algorithm is responsible for finding the operation point where the maximum power from the PV panel can be extracted.

When the batteries are fully charged, a coordinated control strategy is necessary to command the BESS control loop to stop absorbing power and therefore, a new equilibrium point is achieved as expressed in (2):

$$\left.\begin{array}{c} P_{HESS} \approx 0 \\ \sum_{j=1}^{m} P_{LOAD_j} \approx \sum_{i=1}^{n} P_{RES_i} \end{array}\right\} \quad (2)$$

Moreover, grid-following units have to reduce their power contribution to match the loads' consumption, shifting from the maximum power point (MPP). This transition must be done smoothly in order to avoid rough transients.

Upon a sudden change in the load conditions, SESS initially provide the required power due to their faster response capability. Therefore, supercapacitors are used to provide/absorb power during the transients. The difference between the reference of the overall power of the HESS and the transient power managed by the SESS is the power reference of the BESS. This strategy ensures an optimal use of both storage technologies, expanding their useful life. Several strategies can be used to split the power share between the SESS and the BESS of the hybrid system. In DC MGs, the most simple way is to obtain the power reference of the batteries by applying a lowpass filter to the overall power reference as shown in (3) [36,37]. In AC MGs, this step is not as straightforward, as discussed in following sections.

$$\left.\begin{array}{c} P_{BESS}(s) = \frac{\omega_c}{s+\omega_c} \cdot P_{HESS}(s) \\ P_{SESS}(s) = P_{HESS}(s) - P_{BESS}(s) \end{array}\right\} \quad (3)$$

3. PLS Concept Applied to Single-Phase MGs

There are some technical papers in the bibliography where the use of PLS is applied to enable communications between converters in a MG. In some of these works, the PLS is applied to DC MGs [18,20,24] where different control strategies can be found. For instance, in [24], the droop profile varies depending on the PLS frequency. This means there is a continuous injection of a sinusoidal signal into the DC bus. On the other hand, in [18] the control strategy is based not only on a droop control, but on keeping the RES units operating at their MPP while the batteries' SoC is in a safe zone. The moment this SoC is high enough to trigger the PLS, the RESs change to a different operation mode. PLS can also be applied to AC power systems (three-phase ones) as proposed in [21,22,25]. In this particular work, the previous ideas are adapted to be used in single-phase small MGs.

3.1. Selection of the PLS Frequency

Figure 2 shows the basic structure of the MG under study. It consists of a master HESS and two slave PV units. The PLS is generated by the BESS and is measured and filtered in the capacitor of the filter of each RES unit. The PLS is triggered when the SoC of the batteries reaches a certain value. Then, it is broadcasted with a certain frequency that increases as the SoC does. The most appropriate frequency range of the PLS has to be studied in order to avoid interactions with key frequencies like the grid, high frequency harmonics, or the bandwidth of the closed loop control of the PLS. In the particular case of [21] and due to sidebands of the injected signal, a frequency of 90 Hz is selected. However, in our case study, there are unknown line impedances that may affect the propagation of the PLS. Another issue that might disturb this signal is the nature of the loads, i.e., resistive, inductive, and their respective apparent power consumptions. Hence, a frequency analysis of the whole system is required, depending on all these factors and the recommended $f_{PLS_{min}}$ in [21] may not be adequate or at least not the best choice for this particular case.

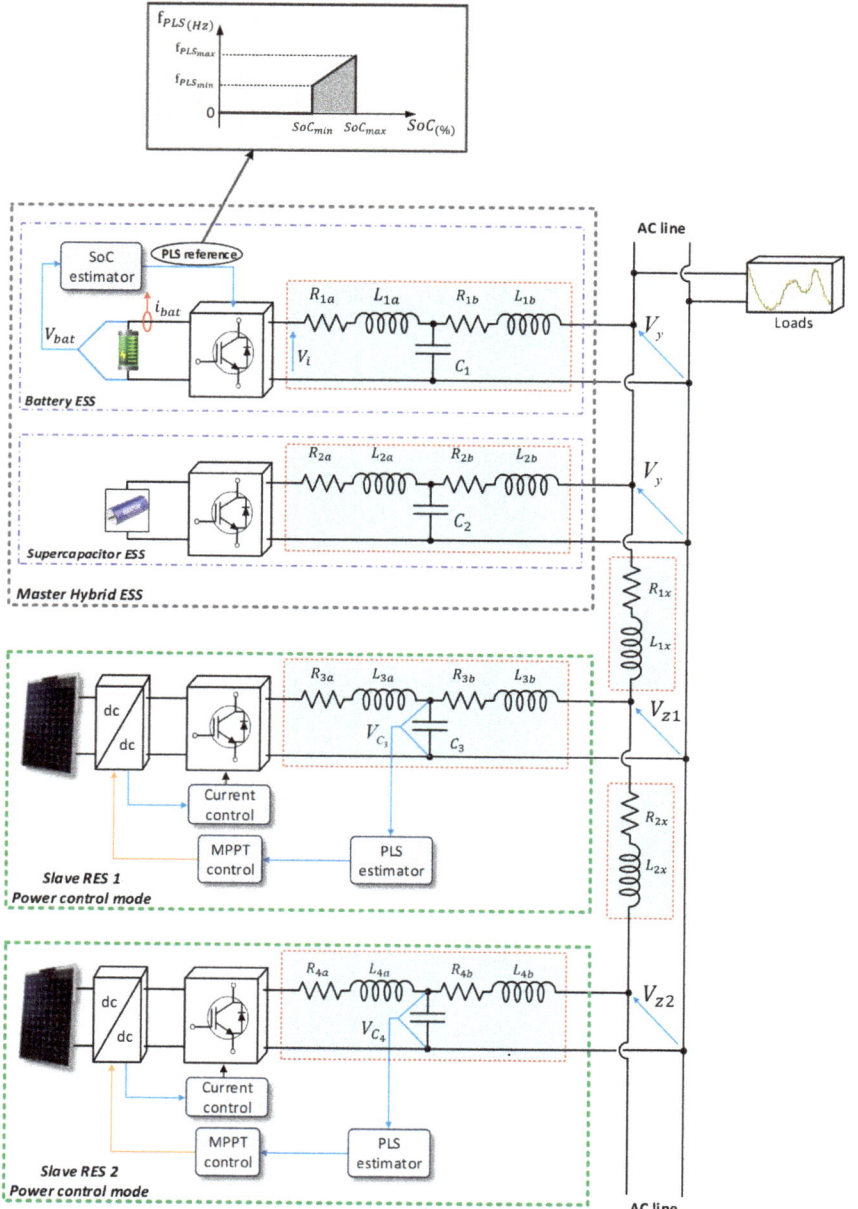

Figure 2. Structure of the AC single-phase microgrid under study.

According to Figure 2, the transfer function between V_{C3} and V_y is denoted by (4):

$$G_{C_3 y}(s) = \frac{V_{C_3}(s)}{V_y(s)} = \frac{1}{(L_{3b}C_3 s^2 + R_{3b}C_3 s) + (L_{1x}C_3 s^2 + R_{1x}C_3 s) + 1} \quad (4)$$

And the transfer function between V_y and V_{C1} is:

$$G_{yC_1}(s) = \frac{V_y(s)}{V_{C_1}(s)} = \frac{Z}{L_{1b}s + R_{1b} + Z} \quad (5)$$

where Z is the impedance of the load.

The relation between the point where the PLS is injected and the voltage at C_1 is therefore:

$$G_{C_1 i}(s) = \frac{V_{C_1}(s)}{V_i(s)} = \frac{1}{L_{1a}C_1 s^2 + R_{1a}C_1 s + 1} \quad (6)$$

Combining (4), (5) and (6), the transfer function between the voltage at the capacitor of the LCL RES 1 filter and V_i is:

$$G_{C_3 i}(s) = \frac{V_{C_3}(s)}{V_i(s)} = G_{C_1 i}(s) \cdot G_{yC_1}(s) \cdot G_{C_3 y}(s) \quad (7)$$

Once the transfer function of the system is known, the effect of the load and line impedances in the PLS needs to be studied. Figure 3 shows the effect of the load over the attenuation of the PLS signal in a Bode diagram. The effect of a pure resistive load is analyzed in Figure 3a, and the consequences of including an inductive performance can be seen Figure 3b. At low to medium range frequencies, the nature of the load has no significant consequences on the Bode diagram.

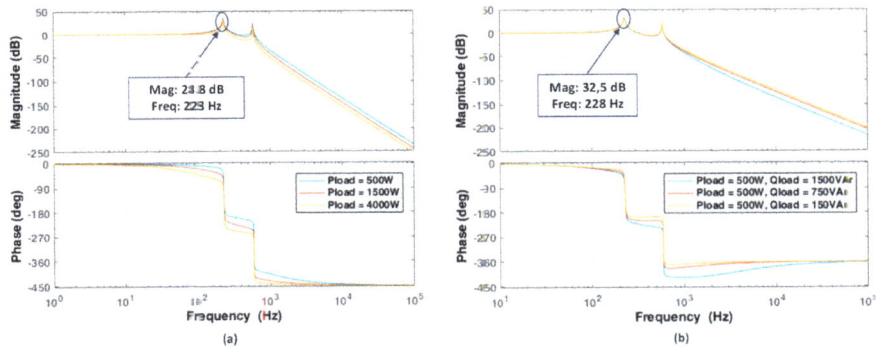

Figure 3. Effect of the load whether it is: (a) resistive (b) inductive.

On the other hand, the effect of line impedances on the PLS attenuation is shown in the Bode diagram of Figure 4. This analysis was performed for several variations in the impedance of the cable.

This very same analysis can be done for the second RES. However, it is necessary to include the corresponding line impedances to obtain the equivalent transfer function as depicted in Figure 5a. The evolution of its root locus is plotted in Figure 5b. The system poles move in different directions depending on those line impedances.

There is a common frequency to all previous analyses that seems suitable to be the PLS one, and that is 228 Hz. This frequency is not affected either by the nature of the load or the line impedances and it is valid for both RES 1 and RES 2 units. This frequency is not close to 100 Hz and it will not interact with the AC loads. Moreover, it is not low enough to disturb the converters' primary control loops. In the particular case of lighting systems, flickering is a key issue. There are some sensitive kinds of lamps (e.g. filament-LED lamps) that could interact with a PLS frequency close to 100 Hz, resulting in undesired situations. That is why the frequency used in [31] is not recommendable. Thus, this value of 228 Hz has been chosen as the center frequency of the power line signal.

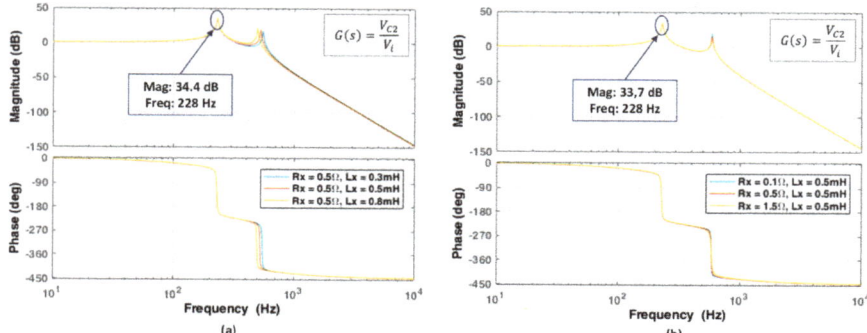

Figure 4. Effect of the line impedances on the power ling signaling (PLS) reception at C_3: (**a**) analysis of the line inductance. (**b**) Analysis of the line resistance.

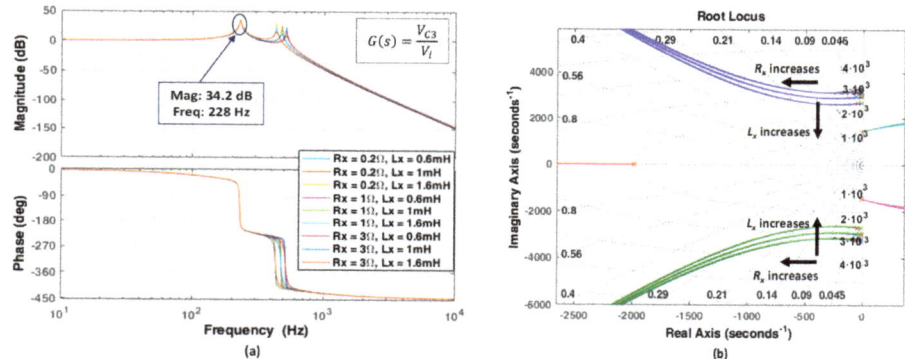

Figure 5. Effect of the line impedances on the PLS reception at C_4: (**a**) Bode diagram. (**b**) root locus.

3.2. Detection of the PLS Frequency

The detection of the PLS is done at the capacitor of the filter of every RES unit attached to the MG. The voltage at these capacitors is measured for two reasons: firstly, to synchronize the grid-following inverter and second, to check if there is any high frequency signal related to communications. As explained in [31], in order to filter possible sidebands, high order filters are required. Choosing a much higher power line frequency signal would have facilitated the filtering process, although it could have interacted with the current loop bandwidth. However, in order to work only with power line signal, two filters are required: a bandstop to attenuate the grid frequency, which has bigger amplitude than the PLS, and a bandpass focused on the objective region (see Figure 6). In this work, Infinite impulse response (IIR) filters have been implemented due to their faster response and fewer coefficients [38].

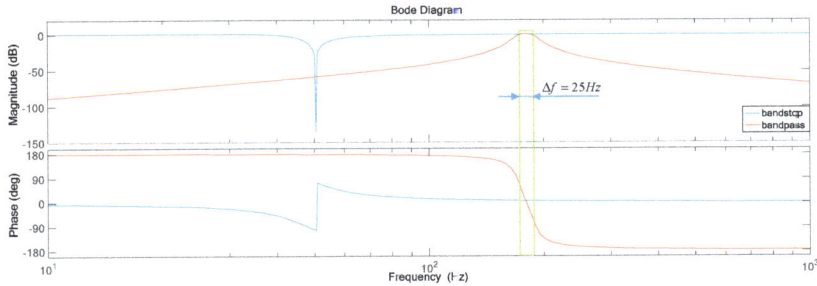

Figure 6. IIR notch and bandpass filters. Bandpass filter order: 4th. Cutting frequencies: 216 Hz, 241 Hz. Sampling rate: 10 kHz.

3.3. PLS Closed-Loop Algorithm

Self-adaptive proportional-resonant (PR) controllers are required for the purpose of ensuring zero steady-state errors of the power line signals. The ideal PR controller has an infinite gain at the AC frequency of ω_{PLS} and no phase shift and gain at other frequencies. By including an anti-windup term and IIR filters, the final control loop of the PLS can be obtained (Figure 7). The PLS signal is generated by a dedicated algorithm that provides the frequency information to the controller, behaving as a self-adaptive one.

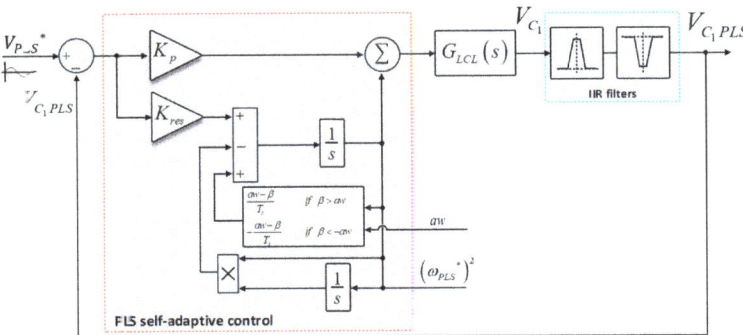

Figure 7. PLS resonant control algorithm.

4. Control Strategy of the MG

The distributed control strategy that prevents batteries from overcharging, as well as maximizing the power contribution of the RESs to the MG, is presented in Figure 8. All different control loops are depicted. A primary control algorithm with two cascade loops for the HESS establishes the MG in a nominal operation point (230 V RMS, 50 Hz). Note how the BESS uses two control loops (voltage and current) to set the MG voltage and frequency, whereas the SESS operates as a grid-following unit, with compensating peak currents. RESs are programmed to inject the maximum power available until power line communications are detected. When that happens, their power contribution is reduced according to the PLS frequency. A simplified flowchart of this algorithm is depicted in Figure 9. This diagram represents the behavior of both HESS and RES units.

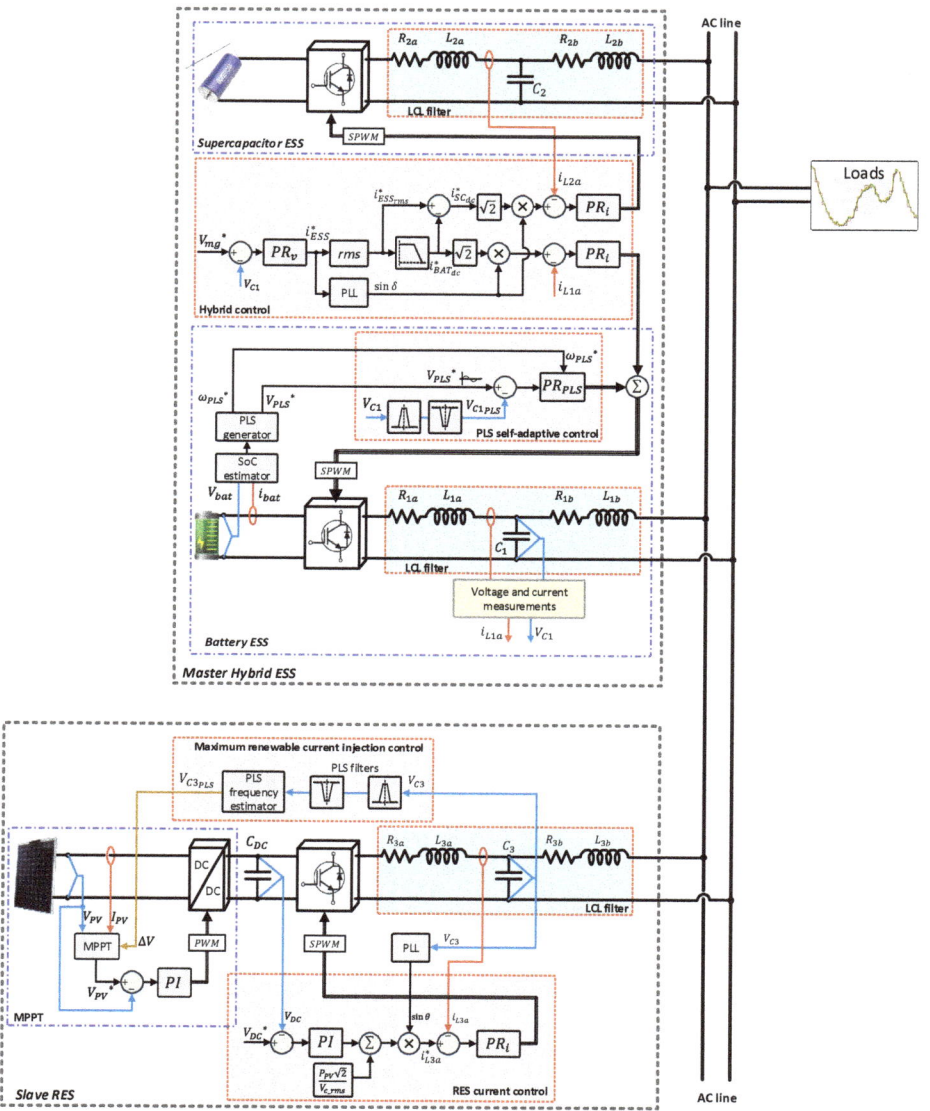

Figure 8. Control diagram of the microgrid (MG).

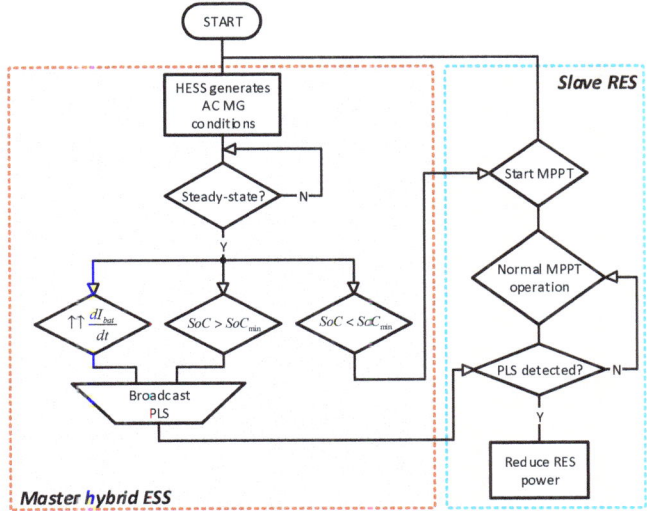

Figure 9. Simplified flowchart of the control algorithm.

4.1. PLS Generation and Event Triggers

The main objective of the PLS communications is the protection of the batteries and therefore, to extend their life. Different events can trigger PLS generation. The primary event that activates communications is a dangerously high SoC of the batteries.

However, there can be other events which could prompt warning signals from the BESS. For instance, a large derivative of the input current into the batteries is not recommended. This could mean there has been an important sudden change in the load or that something is wrong in the MG. When controlling the ScC, the frequency of the PLS is defined by (8):

$$f_{PLS} = \begin{cases} 0 & SoC < SoC_{min} \\ (SoC - SoC_{min}) \cdot m + f_{PLSmin} & SoC_{min} \leq SoC \leq SoC_{max} \end{cases} \quad (8)$$

where m is the slope of the curve depicted in Figure 2.

4.2. PV Slave Unit under Power Control Conditions

The interface of the PV panels can be done in many ways. For instance, in countries with low grid voltages, like Japan, it is becoming very popular to connect the PV panel directly to the grid through an inverter. The MPPT is implemented in the DC-AC inverter [10]. However, the general case is the one depicted in Figure 8, where the MPPT is an independent DC-DC converter that injects the PV current into a DC bus.

This DC link voltage is controlled by the DC-AC inverter. Many MMPT techniques can be found in the bibliography [33–35]. In the present work, a perturb and observe algorithm was used, although it should not be complicated to apply the proposed power control algorithm to any other MPPT technique.

In a safe SoC scenario, the performance of a RES is fixed by the MPPT, providing the maximum available power to the MG. However, if power line communications are detected, the MPPT has to shift the operation point to a different one where the generated power is lower.

The proposed power control algorithm (PLS frequency estimator block in Figure 8) reads the frequency of the PLS and generates a ΔV that shifts the MPP accordingly, generating a new voltage reference to be tracked by the PI controller of the MPPT algorithm. ΔV can be calculated as follows:

$$\Delta V = \begin{cases} 0 & \text{if } f_{PLS} = 0 \\ \left(\dfrac{V_{oc} - V_{MPP}}{f_{PLS_{max}} - f_{PLS_{min}}}\right) \cdot \left(f_{PLS} - f_{PLS_{min}}\right) & \text{if } f_{PLS} \geq f_{PLS_{min}} \end{cases} \quad (9)$$

The previous explanation about shifting the MPP is graphically represented in Figure 10. A red square points out the nominal MPP, i.e., no PLS detected. If there is a PLS broadcast, then the operating point moves to a new one (red circumference), reducing the power obtained from the PV panels. This displacement of the MPP is therefore done according to f_{PLS}.

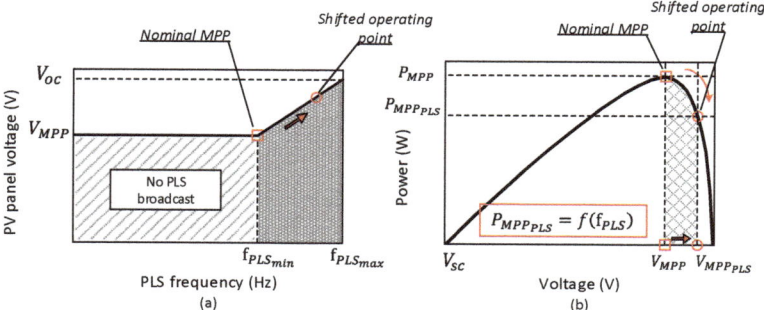

Figure 10. Power control of the photovoltaic (PV) panels when communications are detected. (**a**) PV panel voltage vs. PLS frequency. (**b**) PV power vs. PV voltage.

4.3. Grid-Following Unit (RES Inverter)

The grid-following inverter, which operates together with the MPPT, behaves as a current mode voltage source inverter (CM-VSI). It synchronizes with the MG voltage thanks to a dedicated PLL whereas the current reference is tracked by a single PR controller.

4.4. SoC Estimation

Many *SoC* techniques have been developed during the past few years, allowing users to obtain precise information about remaining battery capacity. Some of these techniques can be found in the bibliography [39–42].

However, these approaches are not easy to reproduce by non-expert researchers. Instead, in this work, a simpler ampere-hour counting method was used to estimate the *SoC* of the batteries:

$$SoC(t) = SoC(0) - \int_0^t \eta_{bat} \dfrac{I_{bat}(t)}{C_{bat}} dt \quad (10)$$

where $SoC(0)$ is the initial *SoC*, C_{bat} is the capacity in Ah, η_{bat} is the charging/discharging efficiency, and I_{bat} is the instantaneous current at the battery [43].

4.5. Plug-and-Play Capability of Additional Units

According to Figure 5, a new RES unit connected to the MG does not affect or damage communications if it is not located far away from the HESS. This unit should be treated the same way as the already-present RES functional units. Therefore, plug-and-play capability can be easily achieved for new RES structures if the proposed control algorithm is adopted.

Nevertheless, in the case of a generic ESS (HESS, BESS, SESS), one should develop a different approach. An interesting strategy would be to operate this second ESS as a backup unit. If the main HESS fails, a second equipment can restore the MG conditions, as an uninterruptible power supply (UPS). There is another option, though, and that is to operate this new unit in parallel with the master HESS, providing a secondary control of frequency and voltage [44], or following a particular droop control [45].

5. Hardware-in-the-Loop Results

The proposed control strategy has been verified through hardware-in-the-loop simulations on a Speedgoat® platform. The parameters of the MG have been gathered in Table 1. One HESS and two PV RES units were simulated supplying different load steps.

Table 1. Power stage and control parameters.

Parameter	Symbol	Value
HESS		
Nominal MG voltage	V_{MG}	230 V
Nominal MG frequency	f_{MG}	50 Hz
Filter inductances	L_a, L_b	1.8 mH
Filter capacitance	C_1, C_2	27 µF
Voltage control inner loop	K_{p_v}, K_{res_v}	0.1, 10
Current control inner loop	K_{p_i}, K_{res_i}	10, 1500
Lower-threshold of SoC	SoC_{min}	95%
Minimum PLS frequency	$f_{PLS\ min}$	226 Hz
Maximum PLS frequency	$f_{PLS\ max}$	231 Hz
PLS control loop	K_{p_PLS}, K_{res_PLS}	5, 250
RES		
Filter inductances	L_a, L_b	1.8 mH
Filter capacitance	C_3, C_4	4 µF
Current control inner loop	K_{p_i}, K_{res_i}	30, 500
Load Steps		
Load 1 ($t = 0s$)	-	1500 W
Load 2 ($t = 55s$)	-	2500 W
Load 3 ($t = 70s$)	-	2500 W
Load 3 ($t = 85s$)	-	−2500 W

These simulation results can be found in Figure 11. According to Figure 9, during the start up of the system, only the HESS is able to supply the loads. Therefore, up to $t = 5\ s$, the batteries are discharging and the PV panels are not operative. Beyond that point, RES units begin to inject power into the MG, and therefore, the power contribution of the BESS is continuously reduced until $t \approx 11\ s$, when the BESS starts to absorb energy and thus, charge the batteries. The SoC of those batteries is climbing under safe values until $t \approx 38\ s$. At that point, it reaches the 95% of its nominal value and PLS is broadcasted. Therefore, the RES units commence to shift the operating point from the MPP, and hence reduce their power contribution.

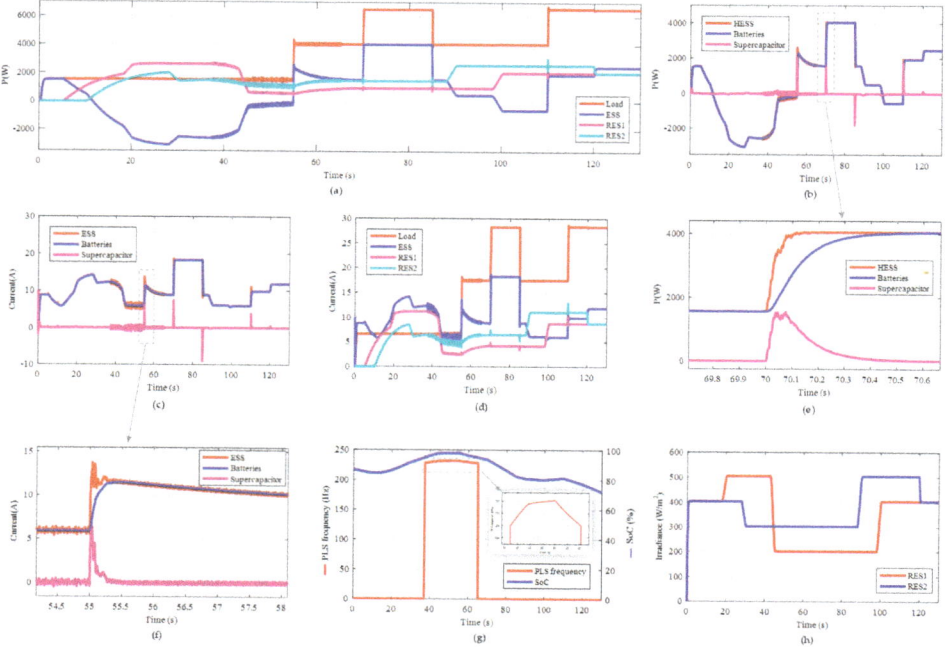

Figure 11. Simulation results: (**a**) Distribution of the powers in the MG. (**b**) Power contribution of the hybrid energy storage system (HESS). (**c**) HESS rms currents. (**d**) Distribution of the rms currents in the MG. (**e**) Detailed view of the HESS powers at $t = 70$ s. (**f**) Detailed view of the HESS rms currents at $t = 55$ s. (**g**) State of charge of the batteries and PLS frequency. (**h**) Irradiance on PV panels.

Three load steps can be found in the simulation results at different moments as summarized in Table 1 (Figure 11a summarizes all these load steps and their effect on the power converters). A new scenario appears at $t = 55$ s when a 2500 W load is connected. The SESS have higher bandwidth and thus it is able to respond faster to this variation. The BESS follows this change of the consumption trend more slowly. Therefore, this transition is assumed by the HESS, while the frequency of the PLS is reduced due to the equivalent *SoC* decrease. The RES are less limited by the PLS, and therefore they will look for a new operation point.

This operating point remains stable until there is a new change in the MG conditions. The *SoC* of the batteries gradually decreases and at $t \approx 65$ s it goes below the SoC_{min}. Power-line communications are thus shut down and RES units again inject all the available power into the MG. This situation remains steady for a few seconds, but then a new load is connected at $t = 70$ s. Again, the SESS is the first unit to react to this transient and the BESS contributes with a softer current reference (Figures 11b and 11e show the power contribution of both SESS and BESS, as well as a more detailed view of the transient at $t = 70$ s).

The MMPT algorithm has been also tested, as can be observed in Figure 11h. The irradiance over both RES units changes during the simulation. The MPPT perfectly tracks the irradiance over the PV panels.

6. Conclusions

In this paper, a distributed control strategy for islanded single-phase microgrids with hybrid energy storage systems based on power line communications has been presented. This approach allows for a coordinated power regulation between the variety of distributed generators and loads that can be encountered in a microgrid. The physical configuration of the microgrid and how to apply power line communications in single-phase islanded microgrids have been studied. The attenuation of this kind of communications can be altered by the integration of more power converters (renewable energy sources, power loads, etc.) and by line impedances, and thus the most suitable range of frequencies for the communications has been calculated.

The proposed control strategy has been validated through hardware-in-the-loop simulations. The renewable energy sources inject power into the microgrid depending on the *SoC* of the batteries of the hybrid energy storage system. When a reference is reached, the grid-following units reduce their power contribution by shifting the operating point in the MPPT algorithm. This displacement of the maximum power point is done according to the frequency of the communication signal. Upon sudden changes in the load conditions, the PV panels again shift their operation point if they are required to.

The calculation of the current references of both systems that make up the hybrid system has been studied. The supercapacitor is responsible for absorbing or delivering the power peaks, while the batteries follow a less aggressive charge/discharge profile. In this way, it is possible to increase its useful life.

Author Contributions: The research study was carried out successfully with contribution from all authors.

Funding: This work has been supported by the Ministry of Economy and Competitiveness of the Government of Spain (MINECO), the Government of the Principality of Asturias and the European Union trough the European Regional Development Fund (ERFD), under Research Grants ENE2013-41491-R (LITCITY Project) and GRUPIN14-076.

Acknowledgments: The authors would like to thank Dan Wu, Nelson Diaz, Lexuan Meng, Juan Carlos Vasquez, Francisco Juarez, Alejandro Suarez, Emilio L. Corominas, Manuel Rico and Pablo Garcia for their valuable contributions.

Conflicts of Interest: The authors declare no conflict of interest.

List of Acronyms

BESS	Battery Energy Storage System
DG	Distributed Generator
ESS	Energy Storage System
HESS	Hybrid Energy Storage System
IRR	Infinite impulse response (filter)
ITS	Intelligent Transfer Switch
MG	Microgrid
MPPT	Maximum Power Point Tracker
PLL	Phase-Locked Loop
PLS	Power-Line Signal
PR	Proportional Resonant (controller)
PV	Photovoltaic (panel)
RES	Renewable Energy Source
SESS	Supercapacitor Energy Storage System
SoC	State of Charge

References

1. Lasseter, R.H. MicroGrids. In Proceedings of the 2002 IEEE Power Engineering Society Winter Meeting, New York, NY, USA, 1 January 2002; pp. 305–308. [CrossRef]
2. Lopes, J.A.P.; Moreira, C.L.; Madureira, A.G. Defining control strategies for microgrids islanded operation. *IEEE Trans. Power Syst.* **2006**, *21*, 916–924. [CrossRef]
3. Guerrero, J.M.; Vasquez, J.C.; Matas, J.; De Vicuña, L.G.; Castilla, M. Hierarchical control of droop-controlled AC and DC microgrids—A general approach toward standardization. *IEEE Trans. Ind. Electron.* **2011**, *58*, 158–172. [CrossRef]
4. Guerrero, J.M.; Chandorkar, M.; Lee, T.; Loh, P.C. Advanced control architectures for intelligent microgrids; Part I: Decentralized and hierarchical control. *IEEE Trans. Ind. Electron.* **2013**, *60*, 1254–1262. [CrossRef]

5. Rocabert, J.; Luna, A.; Blaabjerg, F.; Paper, I. Control of power converters in AC microgrids. *IEEE Trans. Power Electron.* **2012**, *27*, 4734–4749. [CrossRef]
6. Su, W.; Eichi, H.; Zeng, W.; Chow, M.Y. A Survey on the electrification of transportation in a smart grid environment. *IEEE Trans. Ind. Inf.* **2012**, *8*, 1–10. [CrossRef]
7. Blaabjerg, F.; Teodorescu, R.; Liserre, M.; Timbus, A.V. Overview of control and grid synchronization for distributed power generation systems. *IEEE Trans. Ind. Electron.* **2006**, *53*, 1398–1409. [CrossRef]
8. Shahidehpour, M. Don't let the sun go down on PV. *Energy* **2004**, *2*, 40–48. [CrossRef]
9. Koutroulis, E.; Kalaitzakis, K. Design of a maximum power tracking system for wind-energy-conversion applications. *IEEE Trans. Ind. Electron.* **2006**, *53*, 486–494. [CrossRef]
10. Teodorescu, R.; Liserre, M.; Rodriguez, P. Photovoltaic inverter structures. In *Grid Converters for Photovoltaic and Wind Power Systems*; Wiley-IEEE Press: New York, NY, USA, 2011; pp. 5–29. ISBN 978-0-470-05751-3.
11. Rico-Secades, M.; Calleja, A.; Llera, D.G.; Corominas, E.L.; Medina, N.H.; Miranda, J.C. Cosine Phase Droop Control (CPDC) for the dual-active bridge in lighting smart grids applications. In Proceedings of the 2016 IEEE International Conference on Industrial Technology (ICIT), Taipei, China, 14–17 March 2016; pp. 411–418. [CrossRef]
12. Nehrir, M.H.; Wang, C.; Strunz, K.; Aki, H.; Ramakumar, R.; Bing, J.; Miao, Z.; Salameh, Z. A review of hybrid renewable/alternative energy systems for electric power generation: Configurations, control, and applications. *IEEE Trans. Sustain. Energy* **2011**, *2*, 392–403. [CrossRef]
13. Engler, A. Control of parallel operating battery inverters. *Photovolt. Hybrid Power Syst. Conf.* [CD-ROM]. 2000.
14. Stroe, D.I.; Swierczynski, M.; Stroe, A.I.; Laerke, R.; Kjaer, P.C.; Teodorescu, R. Degradation behavior of lithium-ion batteries based on lifetime models and field measured frequency regulation mission profile. *IEEE Trans. Ind. Appl.* **2016**, *52*, 5009–5018. [CrossRef]
15. Zhao, B.; Zhang, X.; Chen, J.; Wang, C.; Guo, L. Operation optimization of standalone microgrids considering lifetime characteristics of battery energy storage system. *IEEE Trans. Sustain. Energy* **2013**, *4*, 934–943. [CrossRef]
16. Fitri, I.; Kim, J.-S.; Song, H. A robust suboptimal current control of an interlink converter for a hybrid AC/DC microgrid. *Energies* **2018**, *11*, 1382. [CrossRef]
17. Asghar, F.; Talha, M.; Kim, S.H. Robust frequency and voltage stability control strategy for standalone AC/DC hybrid microgrid. *Energies* **2017**, *10*, 760. [CrossRef]
18. Pan, H.; Ding, M.; Chen, A.; Bi, R.; Sun, L.; Shi, S. Research on distributed power capacity and site optimization planning of AC/DC hybrid micrograms considering line factors. *Energies* **2018**, *11*, 1930. [CrossRef]
19. Baek, J.; Choi, W.; Chae, S. Distributed control strategy for autonomous operation of hybrid AC/DC microgrid. *Energies* **2017**, *10*, 373. [CrossRef]
20. Both, J. The modern era of aluminum electrolytic capacitors. *IEEE Electr. Insul. Mag.* **2015**, *31*, 24–33. [CrossRef]
21. Rubycon Corporation. Aluminum Electrolytic Capacitor Technical Notes. Available online: http://www.rubycon.co.jp/en/products/alumi/technote5.html (accessed on 1 December 2018).
22. Spanik, P.; Frivaldsky, M.; Kanovsky, A. Life time of the electrolytic capacitors in power applications. In Proceedings of the 12th International Conference ELEKTRO 2014, Rajecke Teplice, Slovakia, 19–20 May 2014; pp. 233–238.
23. Shrivastava, A.; Azarian, M.H.; Pecht, M. Failure of polymer aluminum electrolytic capacitors under elevated temperature humidity environments. *IEEE Trans. Compon. Packag. Manuf. Technol.* **2017**, *7*, 745–750. [CrossRef]
24. Sun, B.; Fan, X.; Qian, C.; Zhang, G. PoF-Simulation-assisted reliability prediction for electrolytic capacitor in LED drivers. *IEEE Trans. Ind. Electron.* **2016**, *63*, 6726–6735. [CrossRef]
25. Byeon, G.; Yoon, T.; Oh, S.; Jang, G. Energy management strategy of the DC distribution system in buildings using the EV service model. *IEEE Trans. Power Electron.* **2013**, *28*, 1544–1554. [CrossRef]
26. Kim, J.Y.; Jeon, J.H.; Kim, S.K.; Cho, C.; Park, J.H.; Kim, H.M.; Nam, K.Y. Cooperative control strategy of energy storage system and microsources for stabilizing the microgrid during islanded operation. *IEEE Trans. Power Electron.* **2010**, *25*, 3037–3048. [CrossRef]

27. Wu, D.; Tang, F.; Dragicevic, T.; Vasquez, J.C.; Guerrero, J.M. A control architecture to coordinate renewable energy sources and energy storage systems in islanded microgrids. *IEEE Trans. Smart Grid* **2015**, *6*, 1156–1166. [CrossRef]
28. Dragievic, T.; Guerrero, J.M.; Vasquez, J.C. A distributed control strategy for coordination of an autonomous LVDC microgrid based on power-line signaling. *IEEE Trans. Ind. Electron.* **2014**, *61*, 3313–3326. [CrossRef]
29. Sun, K.; Zhang, L.; Xing, Y.; Guerrero, J.M. A distributed control strategy based on DC bus signaling for modular photovoltaic generation systems with battery energy storage. *IEEE Trans. Power Electron.* **2011**, *26*, 3032–3045. [CrossRef]
30. Schönberger, J.; Duke, R.; Round, S.D. DC-bus signaling: A distributed control strategy for a hybrid renewable nanogrid. *IEEE Trans. Ind. Electron.* **2006**, *53*, 1453–1460. [CrossRef]
31. Tuladhar, A.; Jin, H.; Unger, T.; Mauch, K. Control of parallel inverters in distributed AC power systems with consideration of line impedance effect. *IEEE Trans. Ind. Appl.* **2000**, *36*, 131–138. [CrossRef]
32. Wu, D.; Tang, F.; Dragicevic, T.; Vasquez, J.C.; Guerrero, J.M. Autonomous active power control for islanded AC microgrids with photovoltaic generation and energy storage system. *IEEE Trans. Energy Convers.* **2014**, *29*, 882–892. [CrossRef]
33. De Brito, M.A.G.; Galotto, L.; Sampaio, L.P.; De Azevedo Melo, G.; Canesin, C.A. Evaluation of the main MPPT techniques for photovoltaic applications. *IEEE Trans. Ind. Electron.* **2013**, *60*, 1156–1167. [CrossRef]
34. Abdelsalam, A.K.; Massoud, A.M.; Ahmed, S.; Enjeti, P.N. High-performance adaptive Perturb and observe MPPT technique for photovoltaic-based microgrids. *IEEE Trans. Power Electron.* **2011**, *26*, 1010–1021. [CrossRef]
35. Subudhi, B.; Pradhan, R. A comparative study on maximum power point tracking techniques for photovoltaic power systems. *IEEE Trans. Sustain. Energy* **2013**, *4*, 89–98. [CrossRef]
36. Wang, C.S.; Li, W.; Wang, Y.F.; Han, F.Q.; Meng, Z.; Li, G.D. An isolated three-port bidirectional DC-DC converter with enlarged ZVS region for HESS applications in DC microgrids. *Energies* **2017**, *10*, 446. [CrossRef]
37. Georgious, R.; Garcia, J.; Garcia, P.; Navarro-Rodriguez, A. A comparison of non-isolated high-gain three-port converters for hybrid energy storage systems. *Energies* **2018**, *11*, 658. [CrossRef]
38. Signal, I.; Magazine, P. What are genetic algorithms? Optimization algorithms. *IEEE Signal Process. Mag.* **1996**, 22–37. [CrossRef]
39. Bhangu, B.S.; Bentley, P.; Stone, D.A.; Bingham, C.M. Nonlinear observers for predicting state-of-charge and state-of-health of lead-acid batteries for hybrid-electric vehicles. *IEEE Trans. Veh. Technol.* **2005**, *54*, 783–794. [CrossRef]
40. Coleman, M.; Lee, C.K.; Zhu, C.; Hurley, W.G. State-of-charge determination from EMF voltage estimation: Using impedance, terminal voltage, and current for lead-acid and lithium-ion batteries. *IEEE Trans. Ind. Electron.* **2007**, *54*, 2550–2557. [CrossRef]
41. Charkhgard, M.; Farrokhi, M. State-of-charge estimation for lithium-ion batteries using neural networks and EKF. *IEEE Trans. Ind Electron.* **2010**, *57*, 4178–4187. [CrossRef]
42. He, H.; Xiong, R.; Zhang, X.; Sun, F.; Fan, J. State-of-charge estimation of the lithium-ion battery using an adaptive extended kalman filter based on an improved thevenin model. *IEEE Trans. Veh. Technol.* **2011**, *60*, 1461–1469. [CrossRef]
43. Díaz, N.L.; Luna, A.C.; Vasquez, J.C.; Guerrero, J.M. Centralized control architecture for coordination of distributed renewable generation and energy storage in islanded AC microgrids. *IEEE Trans. Power Electron.* **2017**, *32*, 5202–5213. [CrossRef]
44. Simpson-Porco, J.; Shafiee, Q.; Dörfler, F.; Vasquez, J.C.; Guerrero, J.M.; Bullo, F. Secondary frequency and voltage control of islanded microgrids via distributed averaging. *IEEE Trans. Ind. Electron.* **2015**, *62*, 7025–7038. [CrossRef]
45. Che, L.; Shahidehpour, M.; Alabdulwahab, A.; Al-Turki, Y. Hierarchical coordination of a community microgrid with AC and DC microgrids. *IEEE Trans. Smart Grid* **2015**, *6*, 3042–3051. [CrossRef]

© 2018 by the authors. Licensee MDPI, Basel, Switzerland. This article is an open access article distributed under the terms and conditions of the Creative Commons Attribution (CC BY) license (http://creativecommons.org/licenses/by/4.0/).

Article

Modeling, Simulation and Analysis of On-Board Hybrid Energy Storage Systems for Railway Applications [†]

Pablo Arboleya [1,*], Islam El-Sayed [1], Bassam Mohamed [1] and Clement Mayet [2]

[1] LEMUR Research Group, Department of Electrical Engineering, University of Oviedo, Campus of Gijón, 33204 Gijón, Spain; islam.aim@gmail.com (I.E.-S.); engbassam@gmail.com (B.M.)

[2] SATIE—UMR CNRS 8020, Conservatoire National des Arts et Métiers, HESAM University, F-75003 Paris, France; clement.mayet@lecnam.net

* Correspondence: arboleyapablo@uniovi.es

[†] This paper is an extended version of our paper published in 2017 IEEE Vehicle Power and Propulsion Conference (VPPC), Belfort, France, 11–14 December 2017, pp. 1–5.

Received: 30 April 2019; Accepted: 6 June 2019; Published: 10 June 2019

Abstract: In this paper, a decoupled model of a train including an on-board hybrid accumulation system is presented to be used in DC traction networks. The train and the accumulation system behavior are modeled separately, and the results are then combined in order to study the effect of the whole system on the traction electrical network. The model is designed specifically to be used with power flow solvers for planning purposes. The validation has been carried out comparing the results with other methods previously developed and also with experimental measurements. A detailed description of the power flow solver is beyond the scope of this work, but it must be remarked that the model must by used with a solver able to cope with the non-linear and non-smooth characteristics of the model. In this specific case, a modified current injection-based power flow solver has been used. The solver is able to incorporate also non-reversible substations, which are the most common devices used currently for feeding DC systems. The effect of the on-board accumulation systems on the network efficiency will be analyzed using different real scenarios.

Keywords: rail transportation power systems; DC power systems; load flow analysis; power systems modeling; load modeling

1. Introduction

The use of power flow algorithms for planning traction networks is a widely-accepted technique [1,2]. However, the use of accurate models of the network and the trains may result in very complicated simulations [3]. In [4], the authors proposed a methodology to perform a fast estimation of the aggregated railway power system and traffic performance. In [5], for instance, the authors tried to reduce the computational burden of the whole simulation, proposing a compression technique that reduces the number of necessary simulations for evaluating the performance of the traction network. In [6], the energetic macroscopic representation (EMR) was used in order to simplify the development of the train and network mathematical models. This approach has been use extensively. For instance, in [7], the authors compared different traction substation models. The models presented using the EMR were extremely accurate and precise [8], and they may describe every single part of the system and are appropriate for applications in which the transient behavior of the trains is critical, such as hardware in the loop applications [9]. A very detailed model of an electric train fed by supercapacitors connected to fuel cells was presented in [10]. Pseudo-static models have been traditionally less accurate, but much faster [11]. They are usually combined with power flow solvers that must be able to manage the complex non-smooth and non-derivable characteristics presented by the new power electronic devices installed within the railways [12].

In this paper, a very simple model for representing a train equipped with a hybrid energy storage system is presented. The combination of regenerative braking with the energy storage system is presented as one of the most effective ways to increase the overall efficiency of traction systems [13–15]. Schedule optimization together with these sets of technologies can be the key to very important energy savings [16–18]. In [19], the savings derived from the installation of an energy storage device in a diesel-electric locomotive were also evaluated. In [20], a method for optimization of the size of the storage devices in order to maximize the energy recovered by the regenerative braking system was presented. An accurate sizing of all the elements in the system was critical for a proper energy management within the system [21–23]. In other cases, the storage systems were installed off-board at the substation level [24]. Many applications, like the one presented in this paper, use hybrid energy storage systems based on batteries and ultracapacitors; usually, the battery covers a base load, and the ultracapacitors are used for feeding the peak power. In order to connect two different storage technologies to the same train, specific power electronic topologies are being investigated, the most common being the DC/DC converter that connects the DC traction network with the storage devices, which uses IGBT technology with a topology that allows the interleaved multi-channel buck-boost operation principle. This converter can adapt the voltage in the energy storage to the catenary voltage. This configuration provides the flexibility required by the operating conditions. Sometimes, series connection of the ultracapacitors or battery packs is needed in order to achieve a specific voltage level. Usually, there are parallel connected IGBT branches that can be fired at the same time or we can shift the firing to reduce the current ripple. In the case of hybrid energy storage systems, there are IGBT branches connected to the battery packs and other IGBT branches connected to the ultracapacitors. Other hybrid topologies use also a common connection; in these cases, the battery packs share the common bus of the ultracapacitors. Usually, buck mode is used for charging the ultracapacitors and the battery. In this situation, the converter drivers control the upper IGBTs, keeping open the devices placed at the bottom of the branch. The current flows from the train DC bus to the energy storage modules. The boost mode (for discharging) uses the IGBTs placed in the bottom, keeping open the ones at the top of the branch. This will produce a current flow from the energy storage devices to the train DC bus. The results obtained using the pseudostatic simulations proposed in this paper are a valuable input for the designers of the energy storage devices. However, a detailed description of the power electronic topology or the storage technology is far beyond the scope of this paper. In order to determine the size of the battery and the ultracapacitors, it is necessary to assess the performance of the system using an ideal integrated energy storage system like the one presented in this paper.

Wayside energy storage can be an alternative to the on-board accumulation systems and in some cases can also be better. In [12,25], the authors mentioned that for choosing the correct infrastructure/train configuration, multiple cases should be analyzed, and all possible technologies should be compared, like for instance reversible substations, wayside energy storage, and on-board energy storage. In order to take advantage of the power regenerated by the trains, the most common solution used in the past was: (1) to optimize the schedule, so when one train is accelerating, another train is braking (this solution is quite difficult to achieve without automatic driving systems); (2) the use of reversible substations; in this case, railway manufacturers are quite reluctant to change the conventional non-controlled diode-based substations because they are inexpensive and robust, so nearly all reversible substations are formed with an IGBT bridge in parallel with the diode bridge; the infrastructure keeps using the diodes to transfer power from AC to DC, and they use the IGBT to put power back into the DC system. Currently, with the drastic reduction in the prices of energy storage systems, it is not clear which is going to be the solution that will prevail, reversibility or accumulation. Probably, accumulation will be a much more reasonable solution in the near future since it helps to maintain the voltages in the DC traction network within the limits, but also increases the overall efficiency of the system. In [25], different configurations were compared, and we can observe that on-board and wayside energy storage are better solutions in terms of efficiency. On-board accumulation is slightly better in terms of efficiency because the losses in the DC system are slightly lower, but wayside energy storage can be more cost-effective since a single accumulation device can be shared by all the trains in the system. Similar conclusions were obtained in [26]; in this work, the use of wayside energy storage improved the overall efficiency by more than 20% and also improved the controllability and the voltage profile of the network. Similar conclusions were obtained in [24,27] supporting the previous conclusions. In [24], the use of energy storage at the substation level reduced the annual cost of energy by 13%. We would like to point out that the main contribution of this paper is the mathematical model; a detailed comparison between the infrastructure alternatives considering wayside and on-board energy storage and reversible substations is beyond the scope of the work, which is focused specifically on providing an accurate and efficient model to simulate on-board energy'storage.

What this work presents is a decoupled model that considers the train and the storage system as separate devices located at the same electrical point and interacting with each other and with the network. The authors did not consider the input filter, and the losses in this device can be estimated and added to the electromechanical and electrochemical device. This approach reduces the complexity of the mathematical model without reducing the accuracy when compared with previous proposed models [28]. The model has been embedded in a commercial software package that uses a modified current injection method to solve the power flow problem [12]; the solver allows combining regenerative trains with diode-based non-reversible substations. It must be remarked that even when there is the possibility of combining regenerative trains with Voltage Source Converter (VSC)-based reversible substations [29,30], in the vast majority of the cases, the existing infrastructures and also the new ones use non-controlled and non-reversible substations due to their robustness, reliability, and low cost [31]. The combination of the two technologies must be done carefully. It should be noticed that without a very precise coordination, the power regenerated by one train could be used to feed another train, but if there is no power demand when the train is regenerating, the voltage in the network will increase, and the substations will be blocked. In such cases, usually the voltage limit is reached, and most of the regenerated power must be burned in the rheostatic braking system. The use of on-board accumulation systems could alleviate this situation [32]. However, it will be demonstrated, the size and effectiveness of the storage system depend highly on the trains' power profile and the schedule.

The results obtained using this kind of simulations are very useful also to study how the accumulation systems are cycled in real applications. There are many on-going research projects that try to predict the life of the accumulation systems depending on the cycling. For instance, the common thought is that load leveling the charge and discharge of the battery packs can contribute to extending the life of the batteries. Recent studies revealed unexpected results. In [33], the authors cycled LiNiCoAl and LiFePO$_4$ modules with 18,650 cells 750 times in a period of six months, and they concluded that the degradation was higher with constant current loads than using dynamic pulse profiles. There is still much research to do in this direction, and every new storage technology requires these kinds of studies. The output of the pseudo-static simulation could be a valuable input for the researchers that try to reproduce real dynamic cycles in accumulation systems.

In the next section, the mathematical model of the train plus the energy storage device in different working modes is explained. Section 3 presents the results obtained using the proposed model in a real traction network. This section has four subsections: in the first three subsections, we describe the feeding infrastructure, the rolling stock, and the proposed scenarios. In the last subsection a deep analysis of the results is presented. Conclusions are stated in Section 4.

2. Mathematical Model of the Train and the Storage System

The proposed train model including the storage system has been simplified in order to lighten the computational burden. In [28], the storage system and the traction equipment of the train were considered as a single element; in this case, the storage system and the traction system separation will substantially simplify the solving procedure. This simplification in the model will not reduce the accuracy because both models are equivalent. The time savings in simulation were around 20%. The solving procedure in this case allows the decoupling between the storage system and the traction equipment, and this is the main reason for this simplification in the mathematical model. From now on, we will use the term train when referring to the traction and breaking device, and storage will be reserved for the accumulation device. Both devices will be modeled as separate, but coupled models.

The storage control will consider the train behavior, as well as the catenary voltage, so we will assume that the train and the storage will be traveling along the grid separated, but connected to the same catenary point.

The joint power and current can be easily calculated as the sum of the current of the train and the storage. For the sign criteria, we will consider that the train will have positive current and power in traction mode when it is absorbing power from the electric grid. In the other case, the storage will have positive power and current when it is in discharging mode. In Figure 1a,b, the schematic representation of the train and storage mathematical models is depicted in traction and braking modes, respectively. We only consider two situations that are the most common ones, the train in traction mode with the storage system discharging and the train in braking mode with the storage system in charging mode.

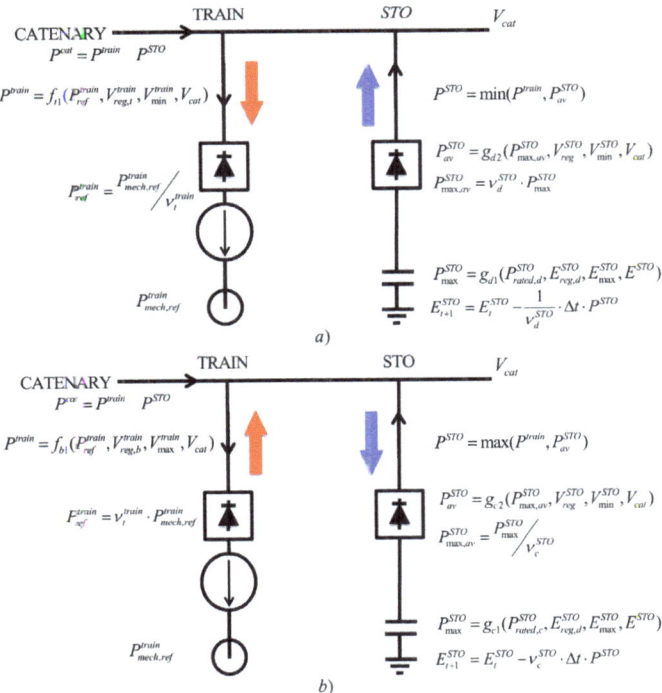

Figure 1. Equivalent mathematical model of the train plus storage system (**a**) Traction mode. (**b**) Braking mode.

2.1. Train Plus Storage Working in Traction Mode

The behavior of the train and the storage in traction mode is summarized in Figure 1a, and the functions f_{t1}, g_{d1}, and g_{d2} are depicted in the curves presented in Figure 2a–c. In the next subsections, a detailed description of the train and the storage model will be presented.

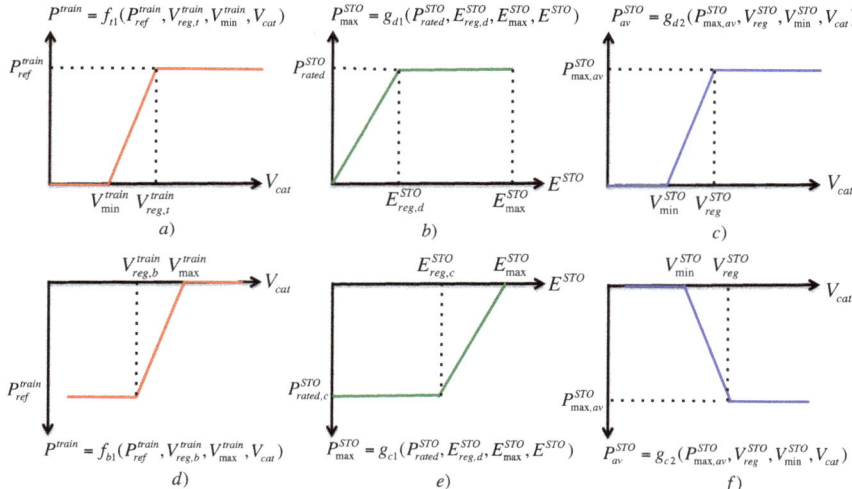

Figure 2. Subfigures (**a–c**) represent the functions that model the behavior of the train and the storage in traction mode. Subfigures (**d–f**) represent the functions that model the behavior of the train and the storage in braking mode. (**a**) Train overcurrent protection in traction mode. (**b**) Maximum power that can be extracted from the storage system in traction mode as a function of the energy level. (**c**) Available discharging power in the storage system as a function of the voltage (overcurrent protection of the storage in traction mode). (**d**) Train squeeze control in braking mode. (**e**) Maximum power that can be injected into the storage system in braking mode as a function of the energy level. (**f**) Available charging power in the storage system as a function of the voltage (overcurrent protection of the storage system in braking mode).

2.1.1. Train Behavior (Positive Power in Traction Mode)

The train model will read first the power reference provided by the external software; we will refer to this power as the mechanical reference power ($P_{mech,ref}^{train}$). It is possible to use coupled models that generate the mechanical power reference during the electrical simulation [34]. However, in most of the cases, the train mechanical model and the electrical problem are solved separately. To obtain the train electrical reference power (P_{ref}^{train}), the mechanical power will be divided by the efficiency of the electromechanical conversion that will include the motors, converters, and the rest of the equipment. This efficiency will be labeled as train efficiency in traction mode (v_t^{train}). This operation can be executed before launching the solver with all positive powers provided by the external software package, so the input file for the solver will already consider this efficiency according to the next expression:

$$P_{ref}^{train} = \frac{P_{mech,ref}^{train}}{v_t^{train}} \tag{1}$$

It must be pointed out that this electric power reference (P_{ref}^{train}) is not the final electric power that the train will demand from the grid (P^{train}). The overcurrent protection can limit the power requested from the grid when the catenary voltage (V_{cat}) is very low for protection purposes.

Each train will have two configuration parameters to set the overcurrent protection ($V_{reg,t}^{train}$ and V_{min}^{train}); in Figure 2a, the function f_{t1} is depicted, and the mathematical expression of such function is expressed in (2).

As can be observed, the actual power consumed by the train depends on the catenary voltage and the train electric power reference, as well as the parameters that define the overcurrent protection ($P^{train} = f_{t1}(P_{ref}^{train}, V_{reg,t}^{train}, V_{min}^{train}, V_{cat})$).

$$P^{train} = \begin{cases} 0 & V_{cat} \leq V_{min}^{train} \\ \frac{V_{cat} - V_{min}^{train}}{V_{reg,t}^{train} - V_{min}^{train}} P_{ref}^{train} & V_{min}^{train} < V_{cat} < V_{reg,t}^{train} \\ P_{ref}^{train} & V_{cat} \geq V_{reg,t}^{train} \end{cases} \quad (2)$$

2.1.2. Storage Behavior (Negative Power in Traction Mode)

The maximum amount of power that can be extracted from the primary active power source of the storage system (ultracapacitors, batteries, flywheel, or whatever other technology used) is labeled as P_{max}^{STO} and depends on the level of charge of the storage system (E^{STO}) with the function g_{d1} (see Figure 2b) that is expressed in (3).

$$P_{max}^{STO} = \begin{cases} P_{rated,d}^{STO} & E^{STO} \leq E_{reg,d}^{STO} \\ \frac{E^{STO}}{E_{reg,d}^{STO}} P_{rated,d}^{STO} & E^{STO} > E_{reg,d}^{STO} \end{cases} \quad (3)$$

The maximum discharging power will be zero if the energy stored is zero, and it will increase in a linear way until a specific energy level is achieved ($\bar{E}_{reg,d}^{STO}$). The maximum discharging power will be constant for the interval between $E_{reg,d}^{STO}$ and the maximum energy that the system can store (E_{max}^{STO}). This maximum power (P_{max}^{STO}) can be calculated at each instant once the storage energy level is updated, but before launching the power flow algorithm, because it is independent of the catenary voltage and the train behavior.

The maximum power available from the storage primary source (P_{max}^{STO}) has to be multiplied by the electrochemical conversion process efficiency in discharging mode (ν_d^{STO}) to obtain the maximum available electric power to be injected into the system ($P_{max,av}^{STO}$). This operation can be also executed before launching the power flow solver.

$$P_{max,av}^{STO} = \nu_d^{STO} P_{max}^{STO} \quad (4)$$

Due to the storage system overcurrent protection, the maximum available power could be constrained if the voltage is too low by means of the function g_{d2} (see Figure 2c). The output of this function is the actual available power that the storage device can inject into the system considering already the catenary voltage constraint (P_{av}^{STO}). To define the function g_{d2}, we need to define also the regulation parameters (V_{reg}^{STO} and V_{min}^{STO}). The analytical expression of the function can be found in (5). Each train can have different regulation parameters. In the case of the storage system, these two parameters will be common for charging and discharging mode.

$$P^{train} = \begin{cases} 0 & V_{cat} \leq V_{min}^{STO} \\ \frac{V_{cat} - V_{min}^{STO}}{V_{reg}^{STO} - V_{min}^{STO}} P_{max,av}^{STO} & V_{min}^{STO} < V_{cat} < V_{reg}^{STO} \\ P_{max,av}^{STO} & V_{cat} \geq V_{reg}^{STO} \end{cases} \quad (5)$$

2.1.3. Coupling between the Train, Storage, and Network

Regarding the interaction between the train, the storage device, and the grid, many operational philosophies can be considered. In this particular case, one of the most popular working modes in real applications has been adopted, giving priority to the energy extraction from the storage system. When the train demands a given amount of power, if this power is available in the accumulation system, it is going to be extracted from it. If not, the train will extract the available power from the storage system, and the rest will be imported from the network. With this control philosophy, the actual power that the storage system is going to inject into the system is going to be calculated following the next expression:

$$P^{STO} = min(P^{train}, P^{STO}_{av}) \tag{6}$$

Due to the employed power flow solving procedure [11], we will use the previous iteration of the catenary voltage to calculate first the power of the train (P^{train}), then we will calculate the storage power (P^{STO}). The catenary net power, representing the whole set (train plus storage device), can be calculated as $P^{cat} = P^{train} - P^{STO}$. Once the catenary power is calculated, a new iteration of the power flow will be launched to obtain a new value of the catenary voltage. The simulation will stop once the difference between all catenary voltages in all nodes in two successive iterations is lower than a specific tolerance.

Once the convergence is achieved, the energy level at the storage system must be updated using the following expression:

$$E^{STO}_{t+1} = E^{STO}_t - (1/v^{ACR}_d) \cdot \Delta t \cdot P^{ACR} \tag{7}$$

2.2. Train Plus Storage Working in Braking Mode

An analogous procedure can be used for describing the train and the storage device working in braking mode. Such behavior is summarized in Figure 1b and the functions f_{b1}, g_{c1}, and g_{c2} depicted in the curves represented in Figure 2d–f. In the next subsections, a detailed description of the train and the storage model will be presented.

2.2.1. Train Behavior (Positive Power in Traction Mode)

Again, the train model will read first the power reference provided by the external software ($P^{train}_{mech,ref}$). To obtain the train electrical reference power (P^{train}_{ref}), the mechanical power will be multiplied by the efficiency of the electromechanical conversion (see Equation (8)). This efficiency will be labeled as the train efficiency in braking mode (v^{train}_b).

$$P^{train}_{ref} = v^{train}_b P^{train}_{mech,ref} \tag{8}$$

The squeeze control can limit the power requested from the grid when the catenary voltage (V_{cat}) is very high for protection purposes. Each train will have two configuration parameters to set the squeeze control ($V^{train}_{reg,b}$ and V^{train}_{max}). In Figure 2d, the function f_{b1} is depicted. The mathematical expression of such a function is expressed in (9). As can be observed, the actual power that the train is going to regenerate depends on the catenary voltage and the train electric power reference, as well as the parameters that define the overcurrent protection ($P^{train} = f_{b1}(P^{train}_{ref}, V^{train}_{reg,b}, V^{train}_{max}, V_{cat}))$.

$$P^{train} = \begin{cases} P^{train}_{ref} & V_{cat} \leq V^{train}_{reg,b} \\ \frac{V_{cat} - V^{train}_{max}}{V^{train}_{reg,t} - V^{train}_{max}} P^{train}_{ref} & V^{train}_{reg,b} < V_{cat} < V^{train}_{max} \\ 0 & V_{cat} \geq V^{train}_{max} \end{cases} \tag{9}$$

2.2.2. Storage Behavior (Negative Power in Braking Mode)

The maximum amount of power that can be injected into the primary active power source of the storage system (P^{STO}_{max}) depends on the level of charge of the storage system (E^{STO}) with the function g_{c1} (see Figure 2e) that is expressed in (10).

$$P^{STO}_{max} = \begin{cases} \frac{E^{STO} - E^{STO}_{max}}{E^{STO}_{reg,c} - E^{STO}_{max}} P^{STO}_{rated,c} & E^{STO} \leq E^{STO}_{reg,d} \\ P^{STO}_{rated,d} & E^{STO} > E^{STO}_{reg,d} \end{cases} \tag{10}$$

Again, this maximum power (P_{max}^{STO}) can be calculated at each instant once the storage energy level is updated, but before launching the power flow algorithm, because it is independent of the catenary voltage and the train behavior.

The maximum power that could be injected into the storage referring to the catenary side of the storage system ($P_{max,av}^{STO}$) has to be calculated using the electrochemical conversion process efficiency in charging mode (v_c^{STC}) according to (11). This operation can be also executed before launching the power flow solver.

$$P_{max,av}^{STO} = \frac{P_{max}^{STO}}{v_d^{STO}} \qquad (11)$$

The overcurrent protection of the storage system is considered also in braking mode using the function g_{c2} (see Figure 2f). The output of this function is the actual available power that the storage device can extract from the system considering already the catenary voltage constraint (P_{av}^{STO}). To define the function g_{c2}, we need to define also the regulation parameters (V_{reg}^{STO} and V_{min}^{STO}). The analytical expression of the function is similar to the one for the traction mode (see Equation (5)).

2.2.3. Coupling between the Train, Storage, and Network

To be consistent with the control philosophy previously described, the system will assign priority to the energy injection into the storage system. When the train regenerates a given amount of power, the storage system will try to use it (if possible) to increase its energy level. If not, the train will inject the maximum allowed power into the storage system, and the rest will be injected into the network. The actual power that the storage system is going to use for charging is going to be calculated following the next expression:

$$P^{STO} = max(P^{train}, P_{av}^{STO}) \qquad (12)$$

The catenary net power, representing the whole set (train plus storage device) can be calculated as $P^{cat} = P^{train} - P^{STC}$. At each instant, the energy level of the storage system can be calculated as follows:

$$E_{t+1}^{STO} = E_t^{STO} - v_c^{STO} \cdot \Delta t \cdot P^{STO} \qquad (13)$$

3. Results' Analysis

In this section, we will analyze in depth the effect of the on-board accumulation system on the trains and network behavior in different scenarios. This section will have four subsections. In the first one, we will describe the basic feeding infrastructure, lines, and substations. In the second subsection, we will describe the rolling stock used in the simulations. The third subsection will be focused on the description of the different scenarios that will be analyzed in Section 4.

3.1. Feeding Infrastructure

The case study in this article will focus on the study of a real network consisting of two lines of 30.84 km and 36.93 km. The voltage level of the system was 3000 V. The simplified diagram of the network can be observed in Figure 3. Line blue is the longest; it had nine stops and four electrical nodes labeled as S1, S2, S3, and S4. The red line shared the first two electrical nodes with the blue line, and it had 17 stops and six electrical nodes labeled as S1, S2, S3, S4, S5, and S6. There were a total of eight electrical nodes and seven lines. Among the eight nodes, only three of them represented feeding substations; the rest were just nodes without any connection with the AC system. The three substations were placed in the nodes S3 and S5 of the red line and in the node S3 of the blue line. The three substations had the same characteristics. All of them were composed by a power transformer with rated power of 3 MW and a short circuit voltage of 5%; the no load output voltage of the rectifier was 3000 V, and the voltage at rated load (1000 A) was 2880 V. The equivalent impedance of each of the three substations in forward mode (AC to DC) was 270 mΩ. The equivalent impedance of the overhead conductor and the rails (return circuit) were respectively 28.605 mΩ/km and 7 mΩ/km. In Table 1, the lengths of the different segments of the red and blue lines can be observed.

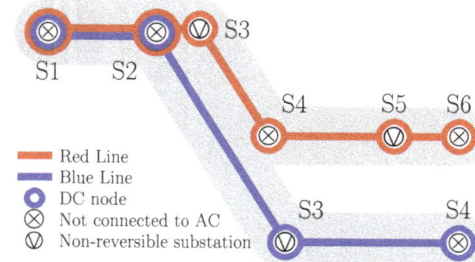

Figure 3. Schematic representation of the case study.

Table 1. Length of the different line segments in km.

	S1 to S2	S2 to S3	S3 to S4	S4 to S5	S5 to S6
Red Line	4.316	0.500	13.800	7.848	4.378
Blue Line	4.316	25.284	7.335	-	-

3.2. Rolling Stock

The train used in both lines was an electrical multiple unit (EMU). The whole unit was 2.940 m wide 4.265 m high and had a total length of 98.05 m, with an unladen weight of 157.3 t. The units were composed of five cars, the two ends having a driver's cabin and a normal floor. The middle car had a normal floor, while the other two cars had a low floor. The five cars were supported on two types of bogies, the trailer bogie and the tractor. The tractor bogie was always shared between two cars. The train was designed to use a standard Iberian track gauge (1668 mm) at a maximum speed of 120 km/h with almost 1000 passengers, although it could reach 160 km/h with minor modifications. The maximum total power of the train was 2.2 MW, and it had regenerative braking. In the base case, the trains were not going to be equipped with on-board energy storage systems. However, we considered the possibility of adding an on-board accumulator system based on a hybrid battery/ultracapacitor technology. The power profiles obtained during the simulation in the accumulation system would help the manufacturer to determine the percentage of energy that must be stored in the battery or ultracap parts. The electromechanical efficiency of the trains in traction and braking modes (v_t^{train}, v_b^{train}), as well as the electrochemical efficiency of the storage system in charging and discharging modes (v_c^{STO}, v_d^{STO}) were set to 0.95. The total accumulation system capacity (E_{max}^{STO}) was 7 kWh, and the on-board energy storage device rated charging and discharging power ($P_{rated,c}^{STO}, P_{rated,d}^{STO}$) was 1 MW.

Regarding the protection curves of the trains and the storage elements, the minimum and the regulation voltage of the train in traction mode ($V_{min}^{train}, V_{reg,t}^{train}$) were set to 1980 V and 2280 V. The same values have been selected for the minimum and regulation voltage of the energy storage system ($V_{min}^{STO}, V_{reg}^{STO}$). In braking mode, the regulation voltage and the maximum voltage of the squeeze control ($V_{max}^{train}, V_{reg,b}^{train}$) were 3300 V and 3600 V, respectively. In the cases in which the on-board accumulation system was activated, the system would be initialized with no charge.

In Table 2, the data summarizing the behavior of the trains in the different trips are collected. In the first column, we can observe the required mechanical power to complete the trip. It can be seen that the slope of the blue line was steeper because the difference between the power required for outward and return journeys was greater than on the red line. The average trip considering the two lines and both directions needed 229 kWh. The mechanical regeneration capacity in Column 2 is the available mechanical power that can be regenerated. Columns 3 and 4 contain the required electrical power and the electrical regeneration capacity considering already the efficiency of the electromechanical conversion. The electrical regeneration capacity was usually around 40% of the required electrical power, except in the S1 to S4 trip of the blue line. In this case, because the train ascended a steep slope, the regeneration capacity was much lower, around 22% of the required electrical power. In the fifth column, we can see the minimum electric consumption. This consumption was calculated as the required electrical power minus the electrical regeneration capacity. Off course, this is a theoretical consumption that considers that we took advantage of all electrical power available for regeneration. This is not true, mainly because of two reasons. First, part of the power that was available to be injected into the catenary was burned in the rheostatic braking system when the squeeze control was activated in order to maintain the catenary voltage below the maximum level. In addition, if the train was equipped with on-board accumulation, the efficiency of the electrochemical conversion during the charging and discharging process also reduced the percentage of available regenerated power that could be reused. For these reasons, we uses these minimum consumption figures as a theoretical ceiling to compare the different solutions, but we must be aware of the fact that we will not reach this theoretical ceiling.

Table 2. Summary of the train behavior in the different trips; all data are in kWh.

Trip	Required Mechanical Energy	Mechanical Regeneration Capacity	Required Electrical Energy	Electrical Regen. Capacity	Min. Electrical Consump. Theoretical
S1 to S6 Red	245	112	258	106	151
S6 to S1 Red	240	107	253	102	151
S1 to S4 Blue	243	61	256	58	198
S4 to S1 Blue	187	99	197	94	103
Average Trip	229	95	241	90	151

3.3. Description of the Selected Scenarios

The authors developed eight different scenarios to study the influence of the accumulation system on the network, as can be observed in Table 3. There were four different paths for the trains, from S1 to S6 and from S6 to S1 for the red line and from S1 to S4 and from S4 to S1 for the blue line. Four different traffic densities were considered. Light traffic scenarios used a train headset of 50 min with 10 departures for each of the above-described routes. The medium traffic scenario considered a train headset of 35 min with 14 departures per route. The dense traffic scenario launched 24 trains per route with a headset of 20 min. Finally, the heavy traffic scenario launched 47 trains per route with a headset of 10 min. The simulation interval was very similar for all scenarios, and it went from eight hours and 18 min for the light traffic scenario to eight hours and 28 min for the heavy traffic scenario. Each of the four traffic densities were simulated without and with on-board accumulation systems without modifying the feeding infrastructure. The obtained results are presented in the next section and will clarify the effect of the on-board accumulation systems on the railway traction networks depending on the density of the traffic.

Table 3. Summary of the different proposed scenarios.

Scenario Code	Traffic Density	On-Board Acc.System	Trains Headset (min)	Trains per Route	Number of Trains	Sim.Time (h:m m)
L0	Light	No	50	10	40	8:18
L1	Light	Yes	50	10	40	8:18
M0	Medium	No	35	14	56	8:23
M1	Medium	Yes	35	14	56	8:23
D0	Dense	No	20	24	96	8:28
D1	Dense	Yes	20	24	96	8:28
H0	Heavy	No	10	47	188	8:28
H1	Heavy	Yes	10	47	188	8:28

3.4. Analysis of the Results

The above-described scenarios were simulated, and the obtained results are analyzed in this subsection. With the software used for simulation, we can obtain time-varying series of each electrical variable in the train or in the feeding network. An example of this detailed analysis can be observed in Figure 4, in which Train Number 4 on the red line route starting from S1 is represented in Scenario L1. It must be noticed how on seven occasions, the train burned part of the regenerated energy using the rheostatic system due to the high catenary voltage, even when the train was equipped with an energy storage system and the storage capacity was not full. It can be observed also how the power extracted/injected into the catenary differed from the train mechanical reference due to the efficiency of the electromechanical conversion process, but also because part of the power was provided by the energy storage system. In Figure 5, the fourth train of the blue line starting from S1 is represented, in this case for the heavy traffic scenario (H1). It has to be remarked that the voltage level was much lower and the network receptivity higher. The burned power in this scenario was nearly zero since the overvoltage protection was only activated on four occasions with a very short duration.

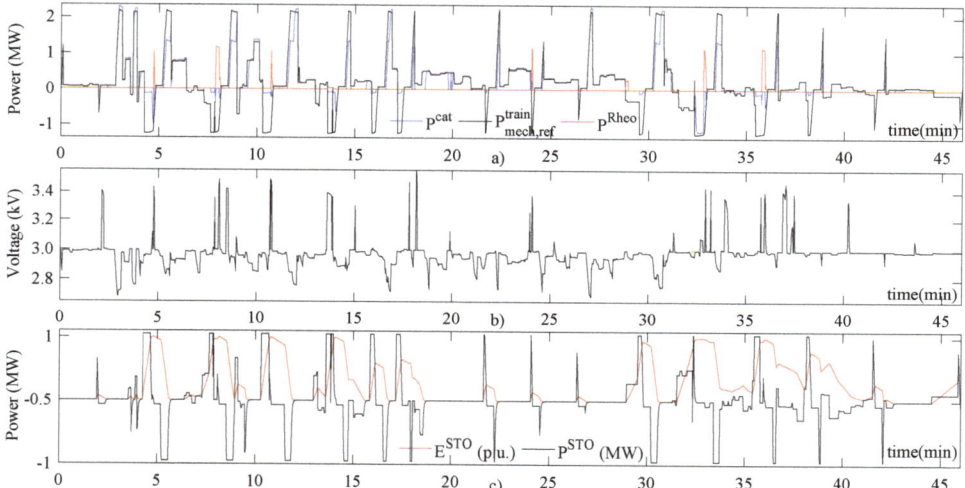

Figure 4. Detailed behavior of the fourth train on the red line from S1 to S6 in the light traffic scenario with on-board energy storage. (**a**) Mechanical reference power; power extracted from the catenary and burned in the rheostatic system. (**b**) Catenary voltage. (**c**) Storage system power (in black) and state of charge of the energy storage system (in red) in (p.u.).

The analysis of the time-varying curves was very interesting since it showed how electrical variables were correlated and it helped to understand the mathematical model proposed in the previous sections. However, in order to determine the effectiveness of the combination of the trains plus the infrastructure, another kind of analysis must be carried out. In Figure 6 is represented the power extracted from the AC network at substation S5 of the red line during the first four hours of simulation for all scenarios. Each subfigure represents two scenarios with the same train headset, but with and without the accumulation system. The correlation between the power extracted from the AC network was higher in the low traffic scenarios. Figure 7 depicts the energy extracted from the AC network at substation S5 in the eight possible scenarios. It must be noticed that the red solid line represents always the scenarios with energy storage. Obviously, scenarios with the same traffic were quite correlated. It should be pointed out that in the light, medium, and dense traffic scenarios, the energy consumption from the AC network in the cases with energy storage systems was a little bit lower. Paradoxically, that is not the case for the heavy traffic scenario in which the energy consumed from the AC network was higher when the trains were equipped with energy storage systems. As will be observed, in heavy traffic scenarios, it was more efficient to share the energy surplus with other trains than storing it in the on-board accumulation system.

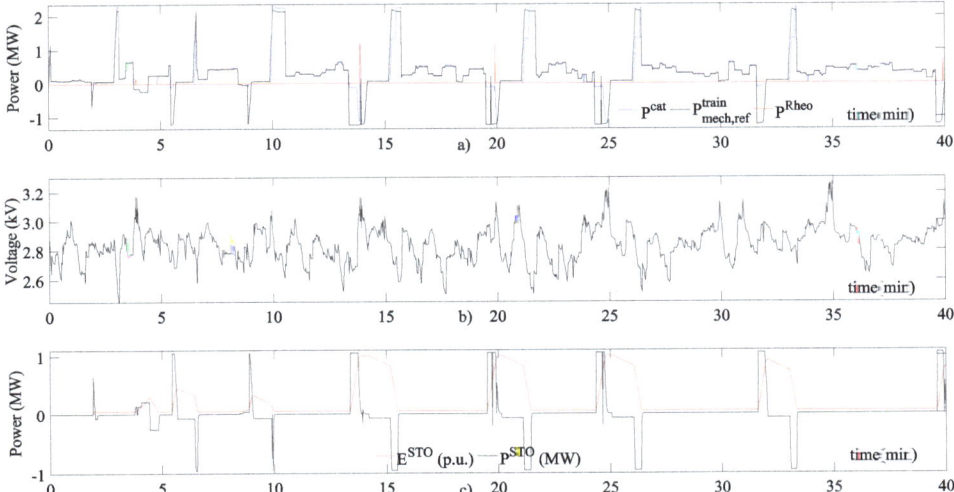

Figure 5. Detailed behavior of the fourth train on the blue line from S1 to S4 in the heavy traffic scenario with on-board energy storage. (**a**) Mechanical reference power; power extracted from the catenary and burned in the rheostatic system. (**b**) Catenary voltage. (**c**) Storage system power (in black) and state of charge of the energy storage system (in red) in (p.u.).

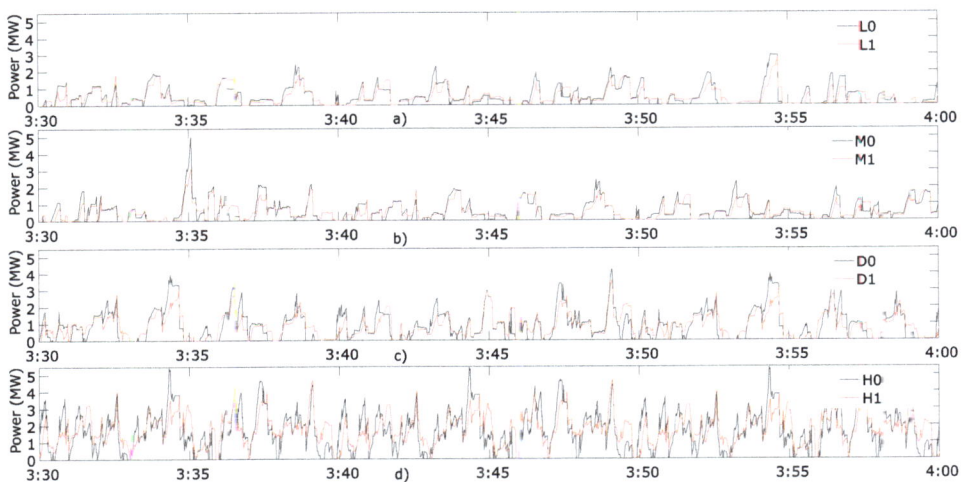

Figure 6. Power in MW obtained from the AC grid by substation S5 on the red line for all scenarios. (**a**) Light Traffic Scenario; (**b**) Medium Traffic Scenario (**c**) Dense Traffic Scenario; (**d**) Heavy Traffic Scenario.

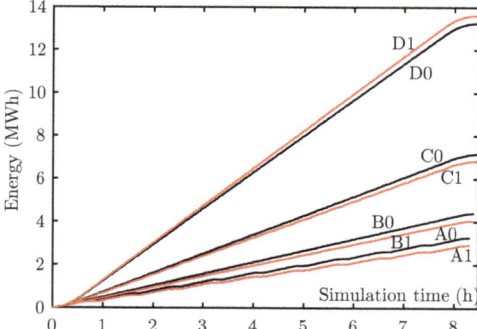

Figure 7. Energy in MWh obtained from the AC grid by substation S5 on the red line for all scenarios.

In Figure 8, we represent the Marey diagrams of each scenario just for the red line. The horizontal axis represents time; in this case, we represented the first 80 min of simulation. In the vertical axis, we represent the position of the train. Solid black lines represent trains from S1 to S6, while dashed-dotted lines represent trains circulating from S6 to S1. Vertical red lines represent the instants at which all the substations were blocked at the same time. It must be noticed that in light, medium, and dense traffic scenarios, the percentage of instants in which all substations were blocked at the same time dropped drastically when we added on-board energy storage systems, improving this percentage by more than 10%. However, in the heavy traffic scenario, the percentage of blocking instants was already very low (5%) in the case without on-board accumulation. The installation of on-board accumulation in the heavy traffic scenario produced the blocking of all substations in only 0.8% of the cases.

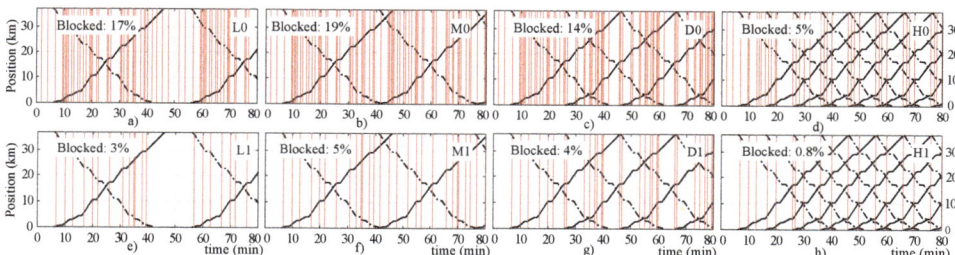

Figure 8. Marey diagrams representing the first 80 min of the schedule of the red line in the different scenarios. Solid lines represent the trains moving from S1 to S6, while dashed-dotted lines represent the trains moving from S6 to S1. The vertical red lines mark the instants at which all the substations in the system were blocked due to the high regenerated power surplus. The number in the top-left corner of each subfigure represents the percentage of instants at which all the substations were blocked at the same time for the whole simulation interval. The scenario is indicated in the top-right corner of each subfigure. (**a**) Light Traffic without accumulation; (**b**) Medium traffic without accumulation; (**c**) Dense traffic without accumulation; (**d**) Heavy traffic without accumulation; (**e**) Light Traffic with accumulation; (**f**) Medium traffic with accumulation; (**g**) Dense traffic with accumulation; (**h**) Heavy traffic with accumulation

In the next paragraphs, we will analyze the aggregated results obtained for the eight scenarios from three points of view. First, we will present a full summary of the system. Second, a summary from the point of view of the trains will be analyzed, and finally, we will show a summary from the point of view of the substations. The use of aggregated data help us to measure the impact of the on-board storage solution in different traffic scenarios for the same feeding infrastructure. In Table 4, we summarize the most representative energies in the system; all the numbers are in MWh. In the first row, we have the electrical energy required by the trains. This energy is going to be the reference energy, since it only depends on the number of trains and does not depend on the equipment of the trains. This energy was the same with and without accumulation. For the light traffic scenario, the 40 trains required a total of 9.62 MWh. For the heavy scenario, the 188 trains required 45.2 MWh, the relation between the number of trains and the required electrical energy being linear. Something similar happened with the second and the third row, which represent the regeneration capacity and the minimum consumption, respectively. The regeneration capacity considers the electromechanical conversion efficiency, and the numbers obtained represent the electrical energy that could be used for injecting into the catenary or charging the storage system. The minimum consumption is a theoretical concept that cannot be reached since it is obtained by subtracting the regenerated capacity from the required electrical energy. Up to now, these numbers only depended on the traffic, and did not vary whether or not the trains had on-board energy storage equipment. The next row (fourth row) represents the energy demanded by the train at the catenary level. It differs from the electrical energy required by the trains for one reason in the case of trains without an accumulation system and two reasons in the case of trains with an on-board accumulation system. In the first case, the train demand was different from the electrical energy requirement because the over-current protection prevented the absorption of too much power in the case of low catenary voltage. In the second case, part of the electrical energy provided by the train can be provided by the on-board accumulation system.

Table 4. Summary of the trains and the network behavior in the different scenarios.

Scenario	Energy (MWh)							
	L0	L1	M0	M1	D0	D1	H0	H1
Req.Electrical	9.62	9.62	13.5	13.5	23.1	23.1	45.2	45.2
Reg.Capacity	3.60	3.60	5.04	5.04	8.64	8.64	16.9	16.2
Min.Comps.	6.03	6.03	8.44	8.44	14.4	14.4	28.3	28.3
Train Demand	8.92	7.23	12.5	10.1	21.8	17.3	43.9	34.8
Train Inject.	1.61	0.55	2.56	0.81	5.3	1.60	13.1	3.99
Trains Net	7.31	6.68	9.94	9.31	16.5	15.7	30.8	30.8
Prov.Subs	7.71	6.98	10.3	9.65	16.7	16.1	31.0	31.2
Rheostatic	2.00	0.49	2.45	0.61	3.36	0.79	3.85	0.84
Non Supp.	0.62	0.16	0.66	0.16	0.51	0.14	0.27	0.09
Grid Losses	0.40	0.31	0.37	0.34	0.19	0.37	0.17	0.40

As can be observed, for the light traffic scenario without accumulation, the trains absorbed 0.72 MWh less than the required energy, so we would have delays in the system. For the heavy traffic scenario, the difference was 1.3 MW. However, in order to compare the scenarios, it is better to compare percentages with respect to the required electrical energy. These percentages can be found in Table 5. The authors are fully aware that this second table is redundant, but we think that it is important to analyze at the same time the data in MWh and in percentage with respect to the required electrical energy. The train demand in the scenario L0 was 92.6% of the required electrical energy, which means that 6.4% of the required energy cannot be provided. In order to distinguish the correlations between the different variables with the different traffic scenarios with and without energy storage and extract conclusions about the trends of the energy savings and consumption, the energetic data in percentage with respect to the total energy demanded by the trains are also represented in Figure 9. As can be observed, we had the same trends as the traffic increased for the cases with and without energy storage. However, even when the trend was the same, the values were very different, and it was necessary to add two different y-scales in order to compare the plots. The real energy demanded by the trains increased with the traffic, as well as the energy injected by the trains. That means that with denser traffic, the number of instants in which the overcurrent and overvoltage protections were activated was lower. We also reduced the energy provided by the substations and the grid losses when we increased the traffic.

Table 5. Summary of the trains and the network in the different scenarios in (%).

Energy in % with Respect to the Electrical Energy Required by Trains								
	L0	L1	M0	M1	D0	D1	H0	H1
Req. Electrical	100	100	100	100	100	100	100	100
Reg. Capacity	37.4	37.4	37.4	37.4	37.4	37.4	37.4	37.4
Min. Comps.	62.6	62.6	62.6	62.6	62.6	62.6	62.6	62.6
Train Demand	92.6	75.1	92.7	75.1	94.4	75.1	97.1	77.0
Train Inject.	16.7	5.7	19.0	6.0	22.9	7.0	29.0	8.8
Trains Net	75.9	69.4	73.7	69.0	71.4	68.1	68.1	68.2
Prov. Subs	80.1	72.5	76.5	71.6	72.3	69.7	68.5	69.1
Rheostatic	20.8	5.1	18.2	4.6	14.5	3.4	8.5	1.9
Non Supp.	6.4	1.7	4.9	1.2	2.2	0.6	0.6	0.2
Grid Losses	4.2	3.2	2.7	2.5	0.8	1.6	0.4	0.9

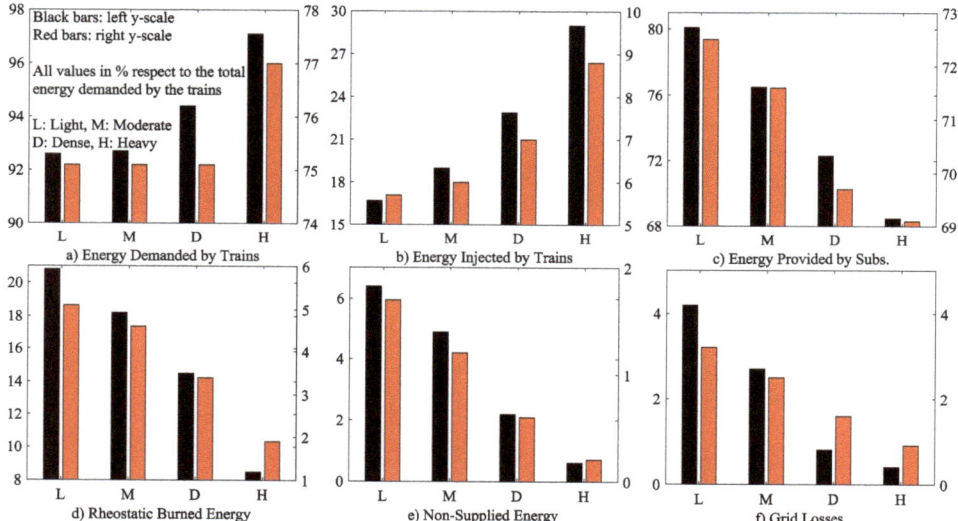

Figure 9. Energy comparison of the different scenarios. In order to compare scenarios with a very different number of trains, all the energy values have been expressed in percentage referring to the total energy demanded by the trains in each scenario. This value can be found for each scenario in the first row of the Table 4. Black bars represent the cases without energy storage, and they are associated with the y-scale on the left. Red bars represent the cases with energy storage, and they are associated with the y-scale on the right.

The non-supplied energy in the case of the heavy traffic scenario without energy storage was only 0.6%. The non-supplied energy is represented in Row 9 of Tables 4 and 5. This non-supplied energy in percentage is a good index of the network congestion, as it can be observed that the more trains in the system, the less the network congestion. In general, we can state that the installation of on-board energy storage always reduced the amount of non-supplied energy. In the worst scenario (light traffic), the non-supplied energy was reduced from 6.4–1.7% when we added storage to the trains. Row 5 represents the energy regenerated by the trains that was actually injected into the feeding system. Obviously, this energy increased with the traffic, but it was always reduced when we added energy storage to the trains. In Row 6, the trains' net energy was obtained subtracting Row 5 (energy injected) from Row 4 (energy demanded). We can see that in the light, medium, and dense traffic scenario, the net energy was always lower when the trains were equipped with energy storage. However, the net energy was nearly the same in the scenarios H0 and H1. Row 7 represents the energy provided by the substations. It is very interesting to remark that even in the worst scenario (L0), the substations provided only 80% of the energy demanded by the trains, the rest being provided by other trains in braking mode. Again, for the light, medium, and dense traffic scenarios, the energy provided by the substations was reduced when we added the energy storage. In the heavy traffic scenario, this trend was inverted, and the energy provided by the substations was slightly higher when we added energy storage to the trains. This is because when we had many trains and they were very close to each other, it was more efficient to use the regenerated energy in other trains than the stored energy in the on-board accumulation system. This is coherent with the data obtained representing the losses in the feeding system (Row 10). In the heavy traffic scenario, the losses were higher when the trains were equipped with an energy storage system. Finally, Row 8 represents the energy burned in the rheostatic system of the train, in all cases, it suffered a significant reduction when we added the on-board accumulation system.

In Table 6, we can observe the trains' average behavior in the different trips and in all scenarios, as well as the average trip considering all possible routes in all scenarios. The table is split into four blocks. In the first one, we represent the total energy obtained from the catenary. It should be noticed that the slope of the blue line was steeper than the slope in the red line. That is the reason why the energy obtained from the catenary in the outward and return journeys was so different, as well as the energy injected into the catenary. It must be noted that for the average trip, the energy obtained from the catenary increased when we increased the traffic because the voltage was more stable (within the limits), and the non-supplied energy decreased. The minimum consumption (theoretical) for the average trip defined in Table 2 was 151 kWh, and this number was obtained subtracting the electrical regeneration capacity from the required electrical energy. The best scenario in terms of average net consumption was H0 (heavy traffic without accumulation), with a net consumption of 164 kWh, only 5.8% above the theoretical consumption. Adding accumulation to the heavy traffic scenario increased the average trip net consumption by 1 kWh. In the rest of the scenarios, adding accumulation always improved the average trip net consumption. For instance, in the light traffic scenario, the net consumption for the average trip was 183 kWh without accumulation and 167 kWh with accumulation. It is very important to remark that when we considered the net consumption in the average trip for comparing with the theoretical limit, we were not considering the non-supplied energy that was quite high in the light, medium, and dense traffic scenarios without accumulation. When we added accumulation to the system, the non-supplied energy in the average trip was always lower than 2% with respect to the required electrical energy for the average trip; and in the case of heavy traffic, lower than 1%.

Table 7 contains the energy analysis of the substations for all scenarios; it is split into three blocks. The first block represents the energy injected into the DC traction system per each substation and the total. In the second block, we can observe the energy injected by each substation per train trip. Finally, this number is compared with the minimum theoretical consumption in Block 3. Again, we can observe that in the scenario H0, the total energy injected into all substations was only 9.3% above the minimum consumption.

Regarding the performance, both models (decoupled and integral) were equivalent from the point of view of the equations, and they provided the same result. However, the time savings on average when simulating the proposed decoupled model were around 20%. On average, the number of iterations of the decoupled model compared to the integral one was higher, from 6.5 iterations (average) to 7.9 iterations (average) per instant. The iterations using the decoupled model were faster. The average time to complete an iteration with the decoupled model was 0.39 ms, while the integral model spent 0.55 ms.

Table 6. Summary of the train behavior in the average trip for all lines and in all scenarios.

Energy Obtained from the Catenary (kWh)								
Trip	L0	L1	M0	M1	D0	D1	H0	H1
S1 to S6 Red	238	183	238	183	245	183	249	186
S6 to S1 Red	243	173	243	173	245	171	251	174
S1 to S4 Blue	230	222	230	221	233	223	246	233
S4 to S1 Blue	181	145	182	146	186	146	189	149
Average trip	223	181	223	181	227	181	234	186
Energy Injected into the Catenary (kWh)								
Trip	L0	L1	M0	M1	D0	D1	H0	H1
S1 to S6 Red	57	16	60	17	67	18	85	23
S6 to S1 Red	38	10	49	10	60	11	70	11
S1 to S4 Blue	9	7	13	7	28	12	34	13
S4 to S1 Blue	57	22	61	24	66	26	90	38
Average trip	40	14	46	15	55	17	70	21
Energy Burned in the Rheostatic Braking System (kWh)								
Trip	L0	L1	M0	M1	D0	D1	H0	H1
S1 to S6 Red	49	12	45	11	40	8	22	4
S6 to S1 Red	64	6	52	5	42	4	31	4
S1 to S4 Blue	50	13	45	12	29	7	24	5
S4 to S1 Blue	37	18	33	16	29	14	5	5
Average trip	50	12	44	11	35	8	21	5
Non-Supplied Energy (kWh)								
Trip	L0	L1	M0	M1	D0	D1	H0	H1
S1 to S6 Red	19	6	19	5	12	5	8	4
S6 to S1 Red	9	2	9	2	7	2	2	1
S1 to S4 Blue	9	4	24	5	21	4	10	2
S4 to S1 Blue	25	4	14	4	11	3	7	2
Average trip	15	4	16	4	13	3	7	2

Table 7. Summary of the substations' behavior in all scenarios.

Energy Injected into the DC Traction System (MWh)								
Substation	L0	L1	M0	M1	D0	D1	H0	H1
S5 (Red line)	2.35	2.03	3.04	2.76	4.88	4.64	8.99	9.02
S3 (Red line)	3.14	2.83	4.22	3.93	6.77	6.54	12.2	12.7
S3 (Blue line)	2.20	2.12	3.04	2.95	5.05	4.93	9.76	9.53
Total	7.71	6.98	10.3	9.65	16.7	16.1	31.0	31.2
Energy Injected into the DC Traction System per Train Trip (kWh/trip)								
Substation	L0	L1	M0	M1	D0	D1	H0	H1
S5 (Red line)	59	51	54	49	51	48	48	48
S3 (Red line)	79	71	75	70	71	68	65	68
S3 (Blue line)	55	53	54	53	53	51	52	51
Total	193	175	184	172	174	168	165	166
Energy Injected into the DC Traction System per Train Trip above the Minimum Theoretical Consumption in %								
	L0	L1	M0	M1	D0	D1	H0	H1
	27.8	15.8	22.1	14.3	15.3	11.3	9.3	10.3

4. Conclusions

A simplified model of a storage system for power flow purposes was presented and tested. The proposed approach considered the train and the accumulation system as different devices placed at the same point of the railway traction network. Both devices were coupled and interacted with each other and with the electrical network. The proposed mathematical model reduced the computational burden of the simulation when it was compared with models that considered the train and the storage in an integral way. The model was embedded in a power flow solver, and extensive results were provided.

Analyzing the results obtained, it can be stated that the behavior of the system when adding this kind of technology cannot be easily predicted. It has been observed how reducing the train headset and increasing the traffic in the system, the non-supplied energy was reduced, as well as the number of instants at which all substations were blocked. In general, we could state that adding on-board accumulation systems to the trains reduced the burned energy in the rheostatic braking system due to the squeeze control activation and also the non-supplied power. However, this effect was highly correlated with the train headset. In order to perform comparisons with different traffic levels, all the energetic data must be normalized using as a rated value the total energy demanded by all trains in each specific scenario. In the cases studied in this work, we observed how the congestion in the network was inversely correlated with the traffic. That means that increasing the traffic, we alleviated the network, and the voltage profile was lower, so the activation of the overvoltage protection was less frequent. We observed also that with the heavy traffic scenario, the trains were able to absorb nearly 98% of the demanded energy when the on-board accumulation was installed and 77% without accumulation. The energy regenerated reached nearly 30% in the heavy traffic scenario with accumulation, when in the same situation without accumulation, it was only 9%, half of the injected regenerated power in the light traffic scenario with accumulation. As a general conclusion, it could be stated that up to a reasonable level, the increasing the traffic can alleviate the system, making it more efficient in relative terms. However, each specific infrastructure can have different behaviors, and it is important to make these kinds of studies during the infrastructure planning stage in order to make the correct decisions.

Author Contributions: Conceptualization, P.A.; methodology, B.M.; software, I.E.-S. and B.M.; validation, C.M. and P.A.; formal analysis, P.A., B.M. and C.M.; investigation, P.A., B.M., I.E.-S., C.M.; resources, P.A.; data curation, B.M. and I.E.-S.; writing—original draft preparation, P.A.; writing—review and editing, P.A. and C.M.; visualization, I.E.-S.; supervision, P.A; project administration, P.A.; funding acquisition, P.A.

Funding: This research was partially funded by CAF TE and the Spanish Ministry of Science, Innovation and Universities through the CDTI strategic program CIEN (National Business Research Consortiums) under the granted project ESTEFI - FUO-EM-155-15

Acknowledgments: The authors would like to thank CAFTurnkey & Engineering, especially Peru Bidaguren and Urtzi Armendariz for their support during the development of these models.

Conflicts of Interest: The authors declare no conflict of interest. CAF TE provided the infrastructure data to perform the study. The Spanish Ministry of Science, Innovation and Universities had no role in the design of the study; in the collection, analyzes, or interpretation of data; in the writing of the manuscript, or in the decision to publish the results.

Glossary

Inputs

$p_{mech,ref}^{train}$	Mechanical power provided by an external software package; it is positive in traction mode and negative in breaking mode.

Parameters

v_t^{train}, v_b^{train}	Electromechanical efficiency of the train in traction and braking modes, respectively.
v_d^{STO}, v_c^{STO}	Electrochemical efficiency of the storage system in discharging and charging modes, respectively.
$E_{reg,d}^{STO}, E_{reg,c}^{STO}$	Storage system regulation energy in discharging and charging modes, respectively, used in the functions g_{d1} and g_{c1}.

E_{max}^{STO}	Maximum amount of energy that can be stored; see functions g_{d1} and g_{c1}.
$V_{reg,t}^{train}, V_{min}^{train}$	Train overcurrent protection parameters in traction mode; see function f_{t1}.
$V_{reg,b}^{train}, V_{max}^{train}$	Parameters for setting up the squeeze control of the train in braking mode; see function f_{b1}.
$V_{reg}^{STO}, V_{min}^{STO}$	Storage system overcurrent protection parameters; see functions g_{c2} and g_{d2}.

Functions

f_{t1}, f_{b1}	Functions for obtaining the actual power of the train in traction and braking mode, respectively; this is the real power absorbed from the grid.
g_{d1}, g_{c1}	Functions for obtaining the maximum amount of power that can be extracted from or injected into the storage system, respectively; this function relates the maximum power in the physical storage system depending on the energy level.
g_{d2}, g_{c2}	Functions for obtaining the available discharging and charging power in the storage system, respectively; this power considers the storage system's overall efficiency and the network conditions.

Variables

E_{STO}	Energy in the storage system at a specific instant.
p^{train}	Actual power that the train exchanges with the electrical traction network.
p^{STO}	Actual power that the storage system exchanges with the electrical traction network.
p^{cat}	Actual power that the set train plus storage system exchange with the electrical traction network; it is positive if it is absorbed from the grid.
p_{ref}^{train}	Train electrical reference power (considering the electromechanical efficiency); this power is obtained directly using the efficiency coefficients from the electrical machines and the input mechanical power; this power includes also the auxiliary power, and it refers to the catenary level.
$p_{rated,c}^{STO}, p_{rated,d}^{STO}$	Energy storage device rated power in charging and discharging modes, respectively.
p_{max}^{STO}	Maximum power at each instant that can be extracted from the storage system; this power is at the accumulation system side; it does not consider the electrochemical efficiency of the process.
$p_{max,av}^{STO}$	Maximum power at each instant that can be extracted from or injected into the storage system; it only depends on the energy stored and the electrochemical efficiency, not on the network conditions.
p_{av}^{STO}	Real available power that the storage system can inject into or extract from the system in discharging and charging mode.
V_{cat}	Catenary voltage.

References

1. Jabr, R.A.; Dzafic, I. Solution of DC Railway Traction Power Flow Systems including Limited Network Receptivity. *IEEE Trans. Power Syst.* **2017**. [CrossRef]
2. Arboleya, P.; Mohamed, B.; Gonzalez-Moran, C.; El-Sayed, I. BFS Algorithm for Voltage-Constrained Meshed DC Traction Networks With Nonsmooth Voltage-Dependent Loads and Generators. *Power Syst. IEEE Trans.* **2016**, *31*, 1526–1536. [CrossRef]
3. Takagi, R.; Amano, T. Optimisation of reference state-of-charge curves for the feed-forward charge/discharge control of energy storage systems on-board DC electric railway vehicles. *IET Electrical Syst. Transp.* **2015**, *5*, 33–42. [CrossRef]
4. Abrahamsson, L.; Söder, L. Fast Estimation of Relations Between Aggregated Train Power System Data and Traffic Performance. *Veh. Technol. IEEE Trans.* **2011**, *60*, 16–29. [CrossRef]

5. López-López, A.; Pecharromán, R.; Fernández-Cardador, A.; Cucala, A. Smart traffic-scenario compressor for the efficient electrical simulation of mass transit systems. *Int. J. Electrical Power Energy Syst.* **2017**, *88*, 150–163. [CrossRef]
6. Mayet, C.; Horrein, L.; Bouscayrol, A.; Delarue, P.; Verhille, J.N.; Chattot, E.; Lemaire-Semail, B. Comparison of Different Models and Simulation Approaches for the Energetic Study of a Subway. *Veh. Technol. IEEE Trans.* **2014**, *63*, 556–565. [CrossRef]
7. Mayet, C.; Delarue, P.; Bouscayrol, A.; Chattot, E.; Verhille, J.N. Comparison of Different EMR-Based Models of Traction Power Substations for Energetic Studies of Subway Lines. *IEEE Trans. Veh. Technol.* **2016**, *65*, 1021–1029. [CrossRef]
8. Mayet, C.; Bouscayrol, A.; Delarue, P.; Chattot, E.; Verhille, J.N. Electrokinematical Simulation for Flexible Energetic Studies of Railway Systems. *IEEE Trans. Ind. Electron.* **2018**, *65*, 3592–3600. [CrossRef]
9. Mayet, C.; Delarue, P.; Bouscayrol, A.; Chattot, E. Hardware-In-the-Loop Simulation of Traction Power Supply for Power Flows Analysis of Multitrain Subway Lines. *IEEE Trans. Veh. Technol.* **2017**, *66*, 5564–5571. [CrossRef]
10. Li, Q.; Wang, T.; Dai, C.; Chen, W.; Ma, L. Power Management Strategy Based on Adaptive Droop Control for a Fuel Cell-Battery-Supercapacitor Hybrid Tramway. *IEEE Trans. Veh. Technol.* **2018**, *67*, 5658–5670. [CrossRef]
11. Mohamed, B.; Arboleya, P.; González-Morán, C. Modified Current Injection Method for Power Flow Analysis in Heavy-Meshed DC Railway Networks With Nonreversible Substations. *IEEE Trans. Veh. Technol.* **2017**, *66*, 7688–7696. [CrossRef]
12. Arboleya, P.; Mohamed, B.; El-Sayed, I. DC Railway Simulation Including Controllable Power Electronic and Energy Storage Devices. *IEEE Trans. Power Syst.* **2018**, *33*, 5319–5329. [CrossRef]
13. Gelman, V. Energy Storage That May Be Too Good to Be True: Comparison Between Wayside Storage and Reversible Thyristor Controlled Rectifiers for Heavy Rail. *Veh. Technol. Mag. IEEE* **2013**, *8*, 70–80. [CrossRef]
14. Gelman, V. Braking energy recuperation. *Veh. Technol. Mag. IEEE* **2009**, *4*, 82–89. [CrossRef]
15. Arboleya, P.; Bidaguren, P.; Armendariz, U. Energy Is On Board: Energy Storage and Other Alternatives in Modern Light Railways. *Electr. Mag. IEEE* **2016**, *4*, 30–41. [CrossRef]
16. Khodaparastan, M.; Dutta, O.; Saleh, M.; Mohamed, A.A. Modeling and Simulation of DC Electric Rail Transit Systems With Wayside Energy Storage. *IEEE Trans. Veh. Technol.* **2019**, *68*, 2218–2228. [CrossRef]
17. Liu, H.; Zhou, M.; Guo, X.; Zhang, Z.; Ning, B.; Tang, T. Timetable Optimization for Regenerative Energy Utilization in Subway Systems. *IEEE Trans. Intell. Transp. Syst.* **2019**. [CrossRef]
18. Liu, P.; Yang, L.; Gao, Z.; Huang, Y.; Li, S.; Gao, Y. Energy-Efficient Train Timetable Optimization in the Subway System with Energy Storage Devices. *IEEE Trans. Intell. Transp. Syst.* **2018**, *19*, 3947–3963. [CrossRef]
19. Mayet, C.; Pouget, J.; Bouscayrol, A.; Lhomme, W. Influence of an Energy Storage System on the Energy Consumption of a Diesel-Electric Locomotive. *Veh. Technol. IEEE Trans.* **2014**, *63*, 1032–1040. [CrossRef]
20. De la Torre, S.; Sánchez-Racero, A.J.; Aguado, J.A.; Reyes, M.; Martínez, O. Optimal Sizing of Energy Storage for Regenerative Braking in Electric Railway Systems. *IEEE Trans. Power Syst.* **2015**, *30*, 1492–1500. [CrossRef]
21. Razik, L.; Berr, N.; Khayyamim, S.; Ponci, F.; Monti, A. REM-S–Railway Energy Management in Real Rail Operation. *IEEE Trans. Veh. Technol.* **2018**. [CrossRef]
22. Lu, S.; Hillmansen, S.; Roberts, C. A Power-Management Strategy for Multiple-Unit Railroad Vehicles. *Veh. Technol. IEEE Trans.* **2011**, *60*, 406–420. [CrossRef]
23. Khayyam, S.; Berr, N.; Razik, L.; Fleck, M.; Ponci, F.; Monti, A. Railway System Energy Management Optimization Demonstrated at Offline and Online Case Studies. *IEEE Trans. Intell. Transp. Syst.* **2018**, *19*, 3570–3583. [CrossRef]
24. Graber, G.; Calderaro, V.; Galdi, V.; Piccolo, A.; Lamedica, R.; Ruvio, A. Techno-economic Sizing of Auxiliary-Battery-Based Substations in DC Railway Systems. *IEEE Trans. Transp. Electr.* **2018**, *4*, 615–625. [CrossRef]
25. Mohamed, B.; El-Sayed, I.; Arboleya, P. DC Railway Infrastructure Simulation Including Energy Storage and Controllable Substations. In Proceedings of the 2018 IEEE Vehicle Power and Propulsion Conference (VPPC), Chicago, IL, USA, 27–30 August 2018; pp. 1–6.
26. Gee, A.M.; Dunn, R.W. Analysis of Trackside Flywheel Energy Storage in Light Rail Systems. *IEEE Trans. Veh. Technol.* **2015**, *64*, 3858–3869. [CrossRef]

27. Alfieri, L.; Battistelli, L.; Pagano, M. Impact on railway infrastructure of wayside energy storage systems for regenerative braking management: A case study on a real Italian railway infrastructure. *IET Electr. Syst. Transp.* 2019. 10.1049/iet-est.2019.0005. [CrossRef]
28. Arboleya, P.; Coto, M.; Gonzalez-Moran, C.; Arregui, R. On board accumulator model for power flow studies in DC traction networks. *Electric Power Syst. Res.* **2014**, *116*, 266–275. [CrossRef]
29. Gomez-Exposito, A.; Mauricio, J.; Maza-Ortega, J. VSC-based MVDC railway electrification system. *IEEE Trans. Power Deliv.* **2014**, *29*, 422–431. [CrossRef]
30. Bai, Y.; Cao, Y.; Yu, Z.; Ho, T.K.; Roberts, C.; Mao, B. Cooperative Control of Metro Trains to Minimize Net Energy Consumption. *IEEE Trans. Intell. Transp. Syst.* **2019**. [CrossRef]
31. Kleftakis, V.A.; Hatziargyriou, N.D. Optimal control of reversible substations and wayside storage devices for voltage stabilization and energy savings in metro railway networks. *IEEE Trans. Transp. Electr.* **2019**. [CrossRef]
32. Clerici, A.; Tironi, E.; Castelli-Dezza, F. Multiport Converters and ESS on 3-kV DC Railway Lines: Case Study for Braking Energy Savings. *IEEE Trans. Ind. Appl.* **2018**, *54*, 2740–2750. [CrossRef]
33. Zhao, J.; Gao, Y.; Guo, J.; Chu, L.; Burke, A.F. Cycle life testing of lithium batteries: The effect of load-leveling. *Int. J. Electrochem. Sci* **2018**, *13*, 1773–1786. [CrossRef]
34. Arboleya, P. Heterogeneous multiscale method for multirate railway traction systems analysis. *IEEE Trans. Intell. Transp. Syst.* **2017**, *18*, 2575–2580. [CrossRef]

© 2019 by the authors. Licensee MDPI, Basel, Switzerland. This article is an open access article distributed under the terms and conditions of the Creative Commons Attribution (CC BY) license (http://creativecommons.org/licenses/by/4.0/).

Article

Application Assessment of Pumped Storage and Lithium-Ion Batteries on Electricity Supply Grid

Macdonald Nko *, S.P. Daniel Chowdhury and Olawale Popoola

Department of Electrical Engineering, Tshwane University of Technology, Pretoria 0183, South Africa
* Correspondence: macconaldnko@gmail.com; Tel.: +27-78-257-5962

Received: 5 May 2019; Accepted: 1 July 2019; Published: 24 July 2019

Abstract: National electricity supply utility in South Africa (Eskom) has been facing challenges to meet load demands in the country. The lack of generation equipment maintenance, increasing load demand and lack of new generation stations has left the country with a shortage of electricity supply that leads to load shedding. As a result, alternative renewable energy is required to supplement the national grid. Photovoltaic (PV) solar generation and wind farms are leading in this regard. Sunlight fluctuates throughout the day, thereby causing irradiation which in turn causes the output of the PV plant to become unstable and unreliable. As a result, storage facilities are required to mitigate challenges that come with the integration of PV into the grid or the use of PV independently, off the grid. The same storage system can also be used to supplement the power supply at night time when there is no sunlight and/or during peak hours when the demand is high. Although storage facilities are already in existence, it is important to research their range, applications, highlight new technologies and identify the best economical solution based on present and future plans. The study investigated an improved economic and technical storage system for generation of clean energy systems using solar/PV plants as the base to supplement the grid. In addition, the research aims to provide utilities with the information required for making storage facilities available with an emphasis on capital cost, implementation, operation and maintenance costs. The study solution is expected to be economical and technically proficient in terms of PV output stabilization and provision of extra capacity during peak times. The research technology's focus includes different storage batteries, pumped storage and other forms of storage such as supercapacitors. The analysis or simulations were carried out using current analytic methods and software, such as HOMER Pro®. The end results provide the power utility in South Africa and abroad with options for energy storage facilities that could stabilise output demand, increase generation capacity and provide backup power. Consumers would have access to power most of the time, thereby reducing generation constraints and eventually the monthly cost of electricity due to renewable energies' accessibility. Increased access to electricity will contribute to socio-economic development in the country. The proposed solution is environmentally friendly and would alleviate the present crisis of load shedding due to the imbalance of high demand to lower generations.

Keywords: pumped storage; solar photovoltaic; lithium-ion batteries; storage; storage operation and maintenance costs; battery management system; state of charge

1. Introduction

South Africa has rich renewable energy (RE) capacity of approximately 4000 MW according the Department of Energy, Eskom has a target of 10,000 GWh to be introduced into the grid. Eskom estimates the photovoltaic (PV) potential to be 64.6 GW. However, just like many other utilities across the world, they are in search of a storage facility for these renewable energies that are expected to supplement the grid during peak hours or whenever required. The dependency on fossil fuels is decreasing while the usage of RE is increasing [1]. This paper presents the research conducted to assess better economic and technical storage systems for clean energy systems using solar/PV plants as a base to supplement the grid. The investigation focused on using lithium-ion (Li) batteries and pumped storage (PPS). Pumped storage is already in use by the utility. This study is intended to provide the utility with information on recent storage battery technologies and how they compare with traditional PPS. Furthermore, if the PPS expected output is achieved, for continuous improvement, new forms of energy storage need to be investigated. Traditionally storage batteries were considered to be very expensive, requiring frequent maintenance and not environmentally friendly in comparison with PPS [2,3]. The study reviews how the development of batteries has improved over the years thereby resulting in cost reduction and traditional perceptions elimination.

The investigation analysis and simulations were carried out using current methods and software called HomerPro® with emphasis on capital cost, implementation, operation and maintenance costs.

To achieve the objectives of the study, which include finding the economic and technical solutions to renewable energy storage, the following design processes were implemented: development of a PV model and its integration with storage batteries using lithium-ion and the design of a pumped storage (PPS) model as well as its integration with the PV model. Comparison of the battery and PPS model was carried out and finally, a developed model arising from the result obtained at different stages of design implementation and efficacy was made. The model development and simulations were carried out in HOMER Pro® environments. Analysis of the results was also carried out to accentuate the impact and contribution of the various scenarios, factors and storage systems. Storage batteries and PPS were simulated and compared, using 3 scenarios: 200 kWh, 600 kWh and 1 MWh storage capacities.

From the results obtained, it can be inferred that both storage systems are significant to PV output stabilisation as required. Both of these can be used to supplement PV in times of high demand and times where PV is not sufficient, due to weather conditions or fluctuations. While these systems have capacities to discharge as and when required, the PPS has a very rapid response and can be used where there are large loads or emergency loading is required. When operated as a hybrid system with the PV, it has been proven that for lower storage requirements, such as the 200 kWh storage bank, PPS is a cheaper option than storage batteries, by operating at almost half the cost. As storage capacity increases to 600 kWh, then the gap closes and PPS becomes approximately 16% cheaper than batteries. The difference reduces as storage capacity increases to 1 MWh. The research has demonstrated that when required storage capacity becomes greater than 1 MWh, then batteries are able to be selected. The assumption was made that both storage resources are available at one site.

Lastly, the research provides the utility with the option to select which storage it requires, depending on the resources at the area where PV is installed. This will improve the stability of the grid, thereby averting outage as a result of load (electricity demand), increasing generation capacity and consumers will have access to power most of the time, which, in the long run, is expected to bring about a reduction in the cost of electricity and contributes to socio-economic development in the country, by creating jobs which will invariably transform into an increase in production capacity. The preferred solution is environmentally friendly and alleviates the present crisis which is a result of high demand and low generation—i.e., supply.

2. Storage Capacity Review

2.1. Lithium-Ion Batteries

The renewable energy output voltage is not constant and requires a storage facility or capacity to regulate the output. In addition, the energy generated can be stored for later use when demand increases, especially for times when sunlight is insufficient to generate power. A simple example is to store energy and release it when the demand is high, such as in the mornings and afternoons. The stored energy can be utilised to supplement the grid. Batteries have the ability to store electrical energy which can be drawn later when required. The other advantages of storage batteries area is that they can assist the solar plant with ramp rate (kW/min) which must be maintained by the solar plant as per utility requirements Ramp rate may be affected by sudden weather changes. This is why batteries are significant: to control the said ramp rate. Mismatches happen between load and generation, causing frequency deviations; however, the storage batteries can assist with frequency control and voltage [4]. The question arising here concerns which batteries are suitable for grid-connected PV plants. There are many types of batteries manufactured but this study only examined the most commonly used ones: lithium-ion and lead-acid batteries. There are important characteristics of batteries that one should consider when choosing the best or most suitable battery solution, such as depth of discharge (DoD), effective capacity, charging/discharging rate and life cycle. In a PV plant, batteries will be connected in parallel-series to meet the required voltage and power.

Lithium-ion batteries are rechargeable type. These batteries over the years have been compared to others in their class; results indicate they currently have lower costs, higher energy, weight less, higher circuit voltage, safe to use, they have an extended life cycle of up to 16 years and better power densities [5]. The efficiency of these batteries has improved over the years to about 70%–99%. More importantly, they no longer require maintenance as it was traditionally [6]. Considering their high cell voltage, they offer few cells required to achieve equal voltage in the same circuit compared to other batteries, resulting in reduced transportation cost and less space required for installation. The fastest charging time of less than 2 h makes them very attractive. Lastly, Li-ion batteries are very friendly and simple to use [7–9]. While these batteries are attractive to use, they require a good strategy, and accurate management to maintain and improve their operating efficiency while enhancing their life span. There are different types of battery management in use while others are being investigated [10–12]. State of charge (SOC) plays an important role in the life span of the batteries, it helps to know when to charge and discharge the batteries. This prevents batteries from damage caused by overcharging and over-discharging. Storage batteries rely on accurate SOC to compete with other forms of storage facilities in terms of offering better storage capacity. Sliding mode observer (SMO), as proposed in [5], focuses on charging and discharging of a battery as exemplified; charging:

$$soc = \hat{V}_{SOC} \tag{1}$$

And discharging;

$$soc = \frac{\int_{t0}^{t1}(I_b^2 R_S + \frac{\hat{V}_f^2}{R_f})d\tau}{\int_{t0}^{t1} V_\bullet I_b d\tau} \tag{2}$$

'^' denotes estimated quantity.

Where $t0$ and $t1$ are starting and end times respectively.

Reading from Figure 1 above; V_{oc} measures the battery open circuit voltage, which is also the function of the battery's State of Charge (SOC). The SOC voltage is represented by V_{soc}, while R_s is the ohmic resistance inserted to control the energy and SOC during battery discharging and charging. RC block (C_f, R_f) shows the battery's reaction upon application of a step load current. The charge capacitor R_{sd} is represented by C_n .where $C_n = 3600 C_Q$ and C_Q is the nominal capacity (A × h). It is to be noted that there are modelling errors, time-varying elements and uncertainties and [5] shows Δf_p, Δf_{oc} and Δf_{soc} can be used to address these errors. To model the battery, the V_{RC} (voltage accross RC) can be given by (3):

$$\dot{V}_{RC} = -\frac{1}{R_f C_f} V_f + \frac{I_b}{C_f} + \Delta f_p \tag{3}$$

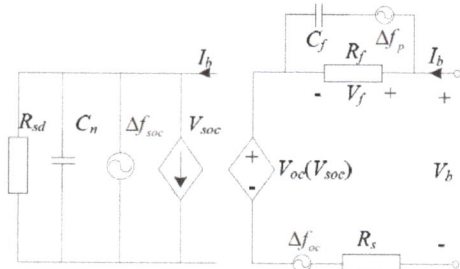

Figure 1. Equivalent circuit of lithium-ion (Li-ion) battery (Courtesy of [5]).

2.2. Pumped Storage

The pumped storage power plant (PPS) has been in existence since 1904; the first installation was used in the 1890s in Italy and Switzerland [13]. PPS built around the world are still functioning, while improved new plants are being continuously built due to their functioning flexibility and adeptness to deliver a quick response to load changes in the system or good electricity price. Synchronous motors are deployed in the system to convert the mechanical energy into electrical energy and, in reverse, to upper reservoirs [14]. PPS is operated by controlling the level of water in the upper reservoir and the output voltage frequency.

Water is stored in the upper tank. During generation mode this water will flow through the hydropower plant to generate electricity. For storage, the used water is pumped from the lower tank into the upper tank using the same reversible turbines. It is noted that some use abandoned mines for lower reservoirs as the height difference between two storage tanks is of paramount importance to generate more mechanical energy [13,14].

Energy is produced by controlling the water level on the upper tank. The Simulink model of water level control is shown in [15]. Traditional PPS used two engines/electric motors, one for pumping water to the upper tank and one to generate electricity through the hydro plant. Recently only one pump can be used to do both works, these pumps are called reversible turbines. These reversible turbines have brought about reduced installation and maintenance costs [16]. Figure 2 below illustrates a typical PPS installation.

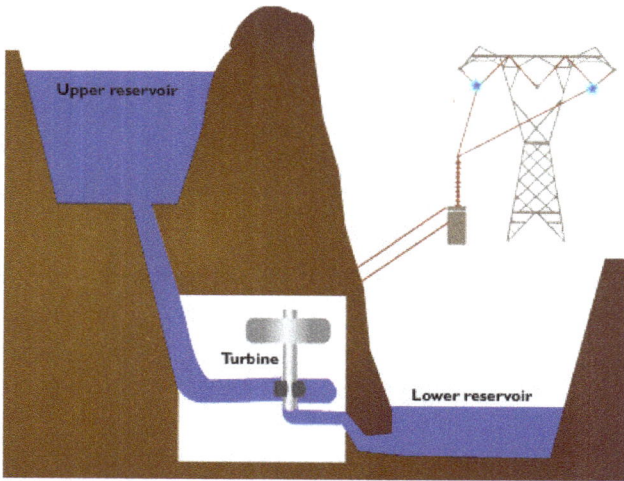

Figure 2. Typical pumped storage plant (PPS) (courtesy of Eskom).

When water flows from the upper reservoir to the discharge reservoir, the motor acts as a turbine and converts gravitational power to mechanical power. The potential power relies on the water head and its flow rate can be expressed as:

$$P_h = \rho g H Q_w \tag{4}$$

where the output mechanical power of the turbine is given by:

$$P_{PT} = \eta \rho g H Q_w \tag{5}$$

P_{PT} is the total output mechanical power from the turbine shaft in Watts, η is the turbine efficiency, ρ is the volume density of water (kg/m^3), g is the gravitational acceleration (m/s^2) due to the height of the upper reservoir. Q_w is the water flow rate passing through the turbine (m^3/s) and H is the effective head of water across the turbine (m) [15].

The efficiency of the pump-turbine is the same in all operating modes, turbine and pump [15,16]. The mathematical expression of the efficiency is expressed below in turbine mode:

$$\eta(\lambda_i, Q_w) = \frac{1}{2}\left[\left(\frac{90}{\lambda_i} + Q_w + 0.78\right)\exp^{(-50/\lambda_i)}\right](3.33 Q_w) \tag{6}$$

with

$$\lambda_i = \left[\frac{1}{\lambda + 0.089} - 0.0035\right]^{-1}, \tag{7}$$

and

$$\lambda = \frac{R A \Omega}{Q_w} \tag{8}$$

R is the turbine radius, A is the Area swept by rotor blades (m^2) and; Ω is the rotational speed (rad/s).

While for fixed water head which is the case for this study, the pump-turbine hydrodynamic torque depends only on water flow rate and rotational speed. The torque equation is given below:

$$T_{PT}(Q_w, \Omega) = \frac{P_{PT}}{\Omega} \tag{9}$$

When the mechanical friction effects are neglected, the pump turbine motion equation is given by:

$$T_{PT}(Q_w, \Omega) - T_{em} = J\left(\frac{d\Omega}{dt}\right) \tag{10}$$

where J is the total inertia of the pump-turbine and motor generator coupling, T_{em} is the motor-generator (electromagnetic) torque.

2.3. Supercapacitors

Another option for supplementing the unreliable solar plant with energy is that of utilising supercapacitors. Supercapacitors are proven to have a long life span, high power density and high dynamics [17]. Supercapacitors are the fastest energy source and can be used to supply the shortage-voltage to achieve DC bus voltage regulation. Supercapacitors store energy between two electrodes in a non-converted electrical form [18].

In the research conducted on [19], the load profile is compared to PV power and the results demonstrate high-low fluctuation in frequency. Supercapacitors are implemented to compensate for low frequencies between the PV plant and the load. These supercapacitors also provide high-frequency components of power and they can absorb high transients due to its rapid response. Supercapacitors have high-density and long life and can last for a life span of 12 years when operated properly [17–19].

This solution of a hybrid supercapacitor with PV plant is mainly used on small scale micro-grids where power requirements are a few megawatts.

It has also been proven that when in the same system or interfaced with batteries' storage (supercapacitor–battery storage combination) the plant is more efficient than when it is PV–battery storage alone [19,20]. These supercapacitors can operate in low and harsh environments, i.e., very high and low temperatures. They are maintenance-free for about 10 years. Unlike the batteries, supercapacitors require less management [21–24].

2.4. Lead Acid Batteries

Lead acid batteries were first invented in the 1800s by the French physician, Gaston Plante and are the oldest known rechargeable batteries. During the discharging mode, both positive and negative plates become lead (II) sulphate PBSO4 and the electrolyte loses much of its dissolved sulphuric acid and becomes primarily water. These batteries are also used in storage requirements facilities, were traditionally used in vehicles and are now found in solar plants. Lead acid batteries can operate under harsher temperature conditions than lithium-ion batteries and furthermore, they provide low-cost storage and are safe to use [8,25]. Lead acid batteries have harmful chemicals which may negatively impact the environment if not disposed correctly. There are several ways of disposing these batteries which is: using a landfill, stabilisation, incineration and recycling [26]. Recycling is becoming a more favourable option because it reduces the environmental impact. Waste lead and acid have serious pollution problems.

2.5. Battery Management System

Battery management system (BMS) improves the life expectancy and operating efficiency of batteries. Each battery type should be managed to ensure that it is not overcharged or undercharged [5,6]. To achieve battery management can be tricky on battery banks. This is because a large number of batteries are connected in series or parallel and all batteries should have equal voltages. BMS becomes effective in this regard. A cell balancing circuit is proposed in [6] as one of the BMS methods to ensure battery banks have equal voltages, although most of these methods are still under study. State of charge (*SOC*) is used to know when to charge or discharge the batteries. An accurate *SOC* will prevent the batteries from overcharging and over discharging, and thus prevent damages. Several methods, such as fuzzy logic, extended Kalman filter, unscented Kalman filter, open circuit voltage, sliding mode observer and non-linear observer, can be used to estimate and improve battery *SOC* [10,11,27–30]. All methods focus on improving the accuracy for *SOC*. BMS predicts the state of health, state of charge and the remaining useful life, besides being used as protection circuit. This is achieved by continuous measurement of battery voltage, current and temperature. This is referred to as a direct method of *SOC* [28,29]. A dual polarisation model is proposed in [29] as the best dynamic performance by providing more accurate *SOC* estimation.

Cost Reduction on Storage Batteries

According to Bloomberg's analysis and predictions in 2017 after surveying more than 50 manufacturers, the cost of lithium-ion batteries was to fall to USD 100/kWh by 2025. The prediction was done when the cost was USD 209/kWh. In 2017, the cost was placed at USD 140/kWh giving indications that the target of USD 100/kWh might be reached by 2020, five years earlier than Bloomberg's prediction; the drop in price being almost 75% [31,32]. Tesla also developed a single battery called a Powerwall system that can store and supply energy for seven days to household applications, assuming the households have solar systems already installed to supply network. In addition, this Powerwall is able to detect grid power loss and spontaneously restore/supply power within milliseconds to a point where outage will not even be noticed. It is indicated that appliances in the houses will continue to function without any interruption. One Powerwall with an estimated usage of 22 kWh/day will cost $6600 inclusive of all materials. More of these technologies are expected and with more funds and time spent on research and development, the battery cost is expected to reduce considerably—i.e., the trend for lithium-ion batteries [33].

Tesla has proved the future of storage lies in batteries and specifically lithium-ion batteries. More research and development should focus on it. It is further suggested in [34] that management of manufacturing process can reduce the cost of batteries. In the same paper, it is suggested that replacing statistical process control with advanced process control and also replacing conventional furnace processing with thermal processing may have an impact in reduction of battery manufacturing costs, as well

2.6. Storage Capacity Review Conclusion

The low initial capital costs, low operational costs, higher life expectancy, user friendly system, high efficiency, high energy and power density are the desired results required to determine the ideal storage facility. It can be deduced that the ideal solution is not possible since no single storage facility can offer all of these good characteristics or outputs [7]. The selection should, therefore, be based on the application and budget. For example, lithium-ion batteries would mostly provide high energy density and have simple operation methods [35]. However, they have a short life span and low power density compared to supercapacitors. This in turn has low-speed response capabilities [18,22]. It is suggested that having supercapacitors in the system can compensate for the shortfalls of batteries because of their high-power density and long-life span [7]. Supercapacitors, however, are complicated to configure on installation, the initial costs are high and not easy to control [22,36]. If choices were to be made on batteries alone, then lithium-ion batteries offer fast charging times and unlike the lead-acid batteries, they require no maintenance [7,35]. Another form of storage in research, in use and under development is the fly-wheel; this stores energy in the form of mechanical energy, is very efficient and has a long-life cycle. Their capability of charging and discharging at high power rates is good; this can be done without loss of efficiency. They are also greenhouse-like pumped storage plants, with no effect on the environment [7]. This research did not focus on this form of storage.

This then gives pumped-storage power plants an advantage over storage batteries because of their functional flexibility and capability to offer quick responses to changes in the system due to a change of loading and/or good price of electricity. Pumped storage, due to its more power output can be used during peak time and thus bring more revenue, a lot of energy can be sold that time and that is the time when energy costs are more expensive [37]. The system can be ready in 90 s and be working on full capacity in 120 s. The switching from pumping to generation or vice versa can happen between 180–240 s, with an efficiency of up to 67% [38] and contribute positively to frequency and voltage control. The only challenge to the pumped storage facility is finding a suitable location based on environmental, topography, geographical and size of reservoirs. There are a lot of initial costs involved and it takes around 10 years to construct and commission. In some areas, reservoirs have to be created, rivers have to be diverted to channel water into the built upper reservoirs [37,38].

3. Investigation Approach and Methodology

The program that was utilised for study costs comparison is Homer Pro®. HOMER (Hybrid Optimisation of Multiple Energy Resources) software was developed to assist engineers with modelling, costs simulations and optimisations of hybrid energy systems. HOMER software can model the grid-tied, off-grid, hybrid systems and stand-alone complex systems. It uses formulas to optimise the required system based on user requirements. It can be used to do cost analysis of different mentioned systems. This can be either renewable energies or traditional grids such as diesel generators, pumped storage, storage batteries, conventional hydro-power plants, boilers, hydrokinetic plant, hydrogen tanks, thermal load controllers and grid [39].

Factors considered and applied during the cost evaluation are stated. The location data such as irradiation, geographical location and temperature were determined using Solaris software (PVplanner) to plot the ideal location for base PV plant. The data obtained was inputted into the Homer Pro®

program. To develop the economical solution using HomerPro® software, the following information was required and inputted: geographical location, meteorological data, hourly/daily/monthly load profiles which the PV system will feed, life span of the system, carbon emissions if any including cost of penalties, cost of each equipment in the system as well as the capital, operation and maintenance. For the cost of each equipment, HomerPro® default and updated values in USD were used. Once all the data is captured, the search space is also used to find the optimal solution for the required load based on the cheapest system analysis to optimise the system behaviour and output. The ultimate solution will then be presented. From [40,41], the levelised costs of energy (COE), annualised cost (Cann) and net present cost (NPC) were determined for each system using below respective formulas/equations; levelised cost:

$$COE = \frac{C_{ann,tot} - C_{boiler} - H_{served}}{E_{served}} \quad (11)$$

where $C_{ann,tot}$ = system annual cost ($/yr); C_{boiler} = boiler marginal cost ($/kWh); H_{served} = total thermal load served (kWh/yr) E_{served} = total electrical load served (kWh/yr).

In the system that does not serve a thermal load then $H_{served} = 0$.

Annualised cost (C_{ann}):

$$C_{ann} = CFR(i, R_{Proj}) \times C_{NPC} \quad (12)$$

where C_{NPC} = the net present cost ($); i = the annual real discount rate (%); R_{Proj} = the project lifetime (yr).

Net present cost (NPC):

$$C_{NPC} = \frac{C_{ann,tot}}{CRF(i, N)} \quad (13)$$

$$CRF(i, N) = \frac{i(1+i)^N}{(1+i)^N - 1} \quad (14)$$

where, $C_{ann,tot}$ is the total annual cost ($/year) which includes the capital, replacement, annual operating and maintenance and fuel costs. CRF is the capital recovery factor, used to calculate the present value of a series of equal annual cash flows, i is the real interest rate (%) and N is the project lifetime (in number of years).

Lastly, it is of great significance to include the operation and maintenance costs as they are a major part of any system including any penalties due to emissions/pollution. However, the system proposed is environmentally friendly and thus no emissions penalties will be incurred. Operation and Maintenace

(O&M) costs will be calculated with the equation using HomerPro®:

$$C_{om} = C_{om,\ fixed} + C_{cs} + C_{emissions} \quad (15)$$

where $C_{om,fixed}$ = system fixed O&M costs ($/yr), C_{cs} = penalty for capacity shortage ($/yr), $C_{emissions}$ = Penalty for emissions ($/yr). Which for this system will be neglected.

For any system to be analysed, the load needs to be connected. The resultant load is based on residential loading. This means each house is already deduced to have usage of 12 kWh per day as calculated below in Table 1. To determine how many houses can be fed with this envisaged demand output, the calculation below is applied.

Assuming a plant of 5 MW *peak* capacity and load factor of 0.2 as given by HomerPro® for residential loading, the *average load* for 5 MW peak will is deduced by:

$$LF = \frac{Average\ Load}{Peak\ Load} \quad (16)$$

$Average\ load\ =\ LF\ \times\ Peak\ Load$
$=\ 0.2\ \times\ 5\ MW$
$=\ 1\ MW$

Average kWh = 1 MW × 24 h, i.e., 24 MWh per day.

Load factor (LF); the load factor is defined as a dimensionless number equal to the *average load* divided by the *peak load*.

Since *per household consumption* is 12 kWh, it can be calculated how may houses can be fed from this plant:

$$\text{Household Connections} = \frac{Average\ Load}{Per\ Household\ Consumption} = \frac{24\ 000\ kWh}{12 kWh} = 2000, \quad (17)$$

two thousand houses can be fed from a 5 MW plant. Figure 3 below shows the typical daily and seasonal profile of a residential customer in South Africa.

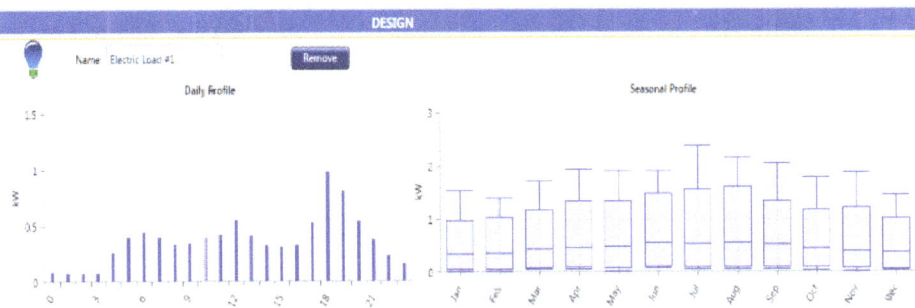

Figure 3. Daily and seasonal profile per household.

4. Simulation, Results and Discussion with Investigation

PPS and Li-ion batteries were modelled using six different scenarios on HomerPro® with system loading of 1 MW, 3 MW and 5 MW. Each system load was to have storage of 20% capacity for PPS and Li-Ion. An off-grid system is applied for all investigations.

The below case studies were simulated and analysed:

- **Case study 1**: 5 MW Load, PV Plant and 20% Storage Battery Capacity.
- **Case study 2**: 5 MW Load, PV Plant and 20% Pump Storage Plant.
- **Case study 3**: 3 MW Load, PV Plant and 20% Storage Battery Capacity.
- **Case study 4**: 3 MW Load, PV Plant and 20% Pump Storage Plant.
- **Case study 5**: 1 MW Load, PV Plant and 20% Storage Battery Capacity.
- **Case study 6**: 1 MW Load, PV Plant and 20% Pump Storage Plant.

Throughout the discussion that follows in the case studies, the load varied per case study and the PV plant was kept constant. Just the storage facilities were varied to determine the most economical solution. This study considered only Li-ion type batteries.

Case Study 1: 5 MW Load, Photovoltaic (PV) Plant and 20% Storage Battery Capacity

The peak load was 5 MWh, and merely 20% of the load, which was just 1 MWh, had to be stored in Li-Ion batteries. Figure 4 shows the schematic diagram of case study 1. The properties of the storage batteries are provided below in Figure 5. A total of 1000 batteries were required to make 1 MWh storage.

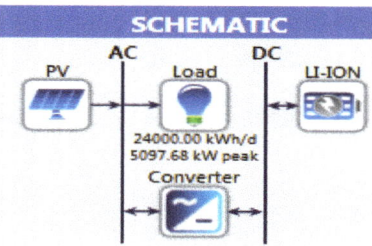

Figure 4. Scenario 1–5 MW photovoltaic (PV) plant, Li-ion batteries, and load connected—off-grid.

Looking at Figures 6–8 below makes it clear that PV would only supply the load at times where there is sun available. This is usually from 7 am. till 5 pm. South African Time. At night time there will be no power supplied to the load and the output is not smooth during the day due to the inconsistency of the weather. This is the reason why storage is required. Figure 8 shows the typical behaviour of a PV plant in the month of January, most importantly the variation of AC load per day while the supply remains almost constant. PV will never be able to supply the load hence storage can be used to at least supplement the PV where and when the load has peaked.

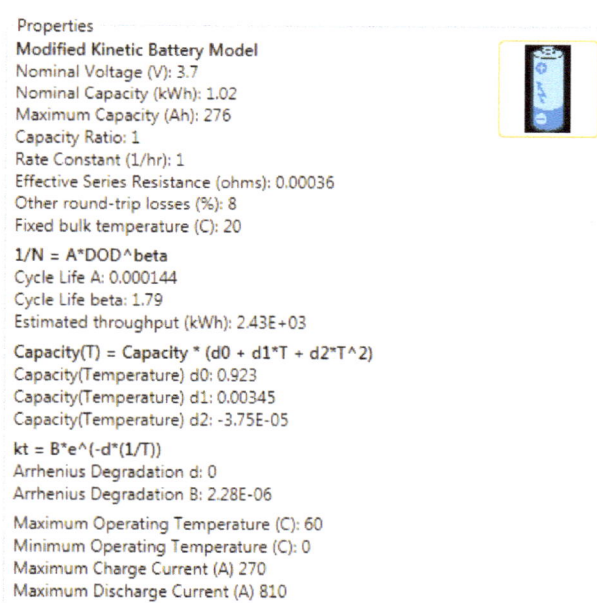

Figure 5. 1 MWh Li-ion battery bank properties.

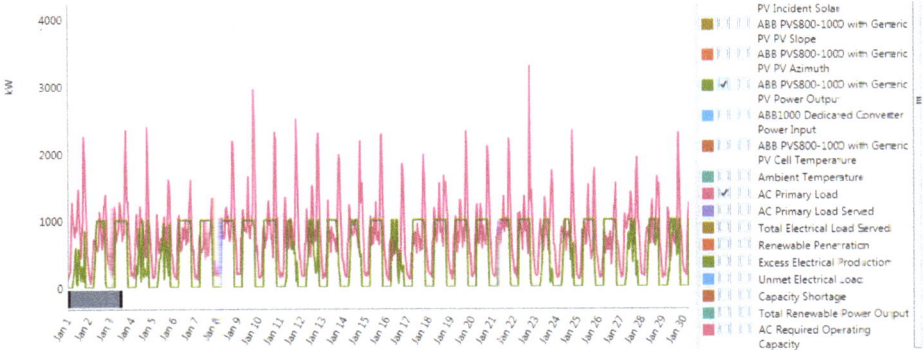

Figure 6. 5 MW PV plant performance supplying load without storage.

Continuing from Figure 6, now the storage battery bank is introduced in the system and that is shown in Figure 9. PV power output is drawn as the green graph and load is the purple one. In Figure 6; where there is no storage facility on the system, PV output starts increasing from 5 am and peaks at 7 am. From there the output is smooth; it is up and down depending on the irradiation on the day. This output started dropping at 18:00 because it depends solely on the sun. From 19:00 little power can be supplied and there will be no power by 20:00. The simulations and results were based on Cape Town times, which are different from other areas because it has longer sun day by nature.

Figure 9 Where the system had a battery storage facility, the PV output behaved the same. The difference came about at 8 am where there was sudden peak loading until 11 am, then storage power could be discharged from batteries to supplement the PV and the same occurred again from 13:00 to 18:00. For example at 18:00, 218 kW could be drawn from batteries when the load was at peak and PV was dropping. Normally, when load drops occur, then PV is used to charge the batteries. Figure 10 shows the 5 MW PV plant net present cost without storage using ABB PSV800-1000 solar panels.

Figure 7. 5 MW PV plant output without storage hourly basis.

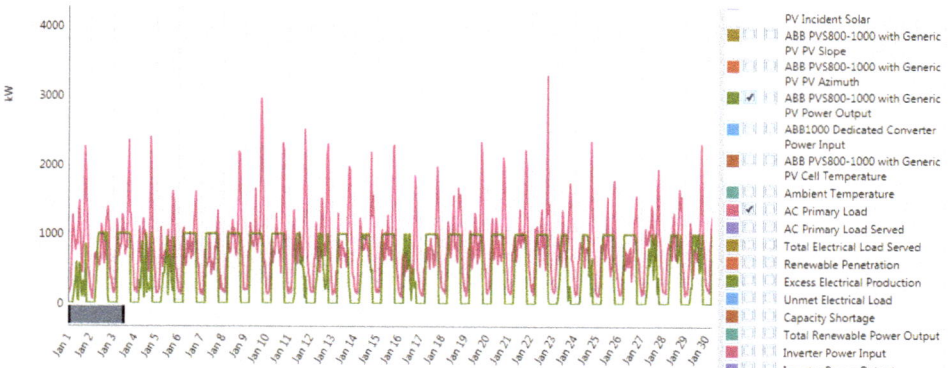

Figure 8. 5 MW PV plant output without storage battery month preview.

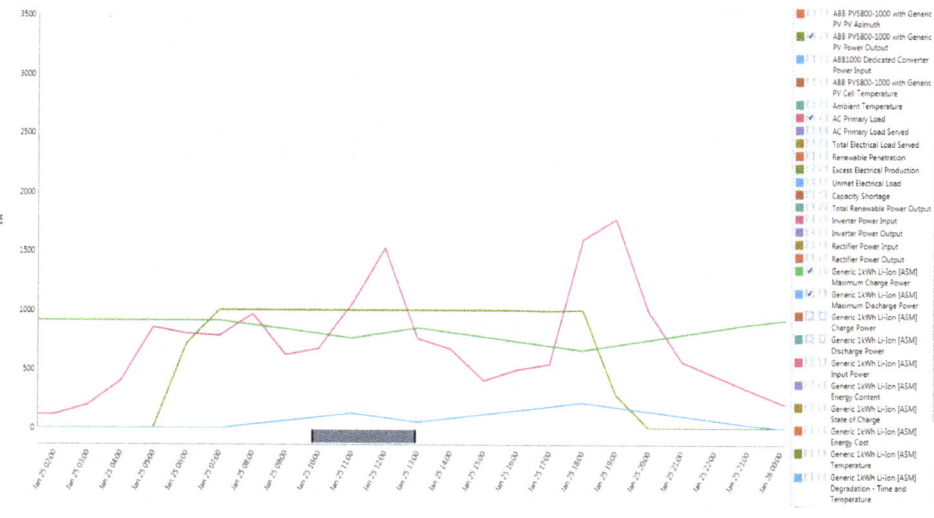

Figure 9. 5 MW PV and Li-ion output on summer day.

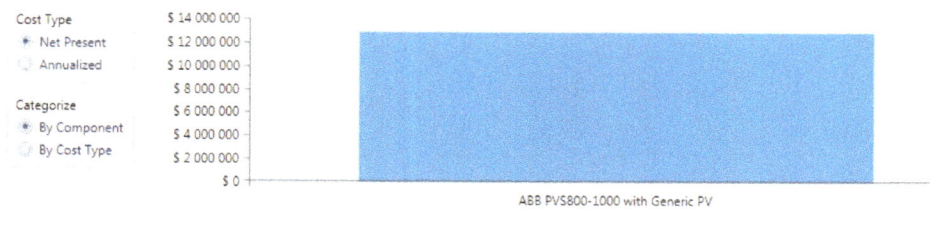

Figure 10. 5 MW PV plant net present cost without storage using ABB PSV800-1000 solar panels.

4.1. Photovoltaic (PV)–Battery Plant Off-Grid Cost Analysis

Based on the results on Figures 10–12, having a PV-Battery power plant in the Western Cape would cost approximately Net USD 15 Million with annual cost of USD 1.4 Million for a project with a life cycle of 25 years. This system is technically excellent but it may require a huge amount of capital to build. This system is 100% renewable with 20% battery storage capacity, 56% unmet load and up to

90% capacity shortage settings. However, the return on investment is very low −1.7%. This then may be too expensive to construct. However, it can be used as a base for smaller systems/plants. Since the system is off-grid, the number of batteries was increased which increases the system cost. Accordingly, it was demonstrated that there would be no investment in building such a system since it will not bring fruitful returns.

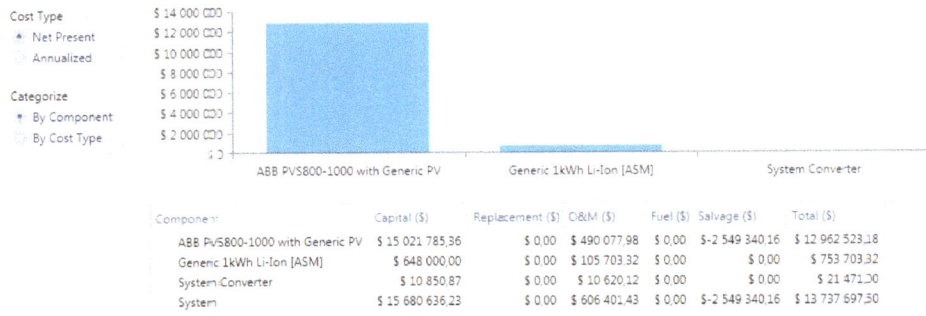

Figure 11. 5 MW PV plant net present cost—batteries.

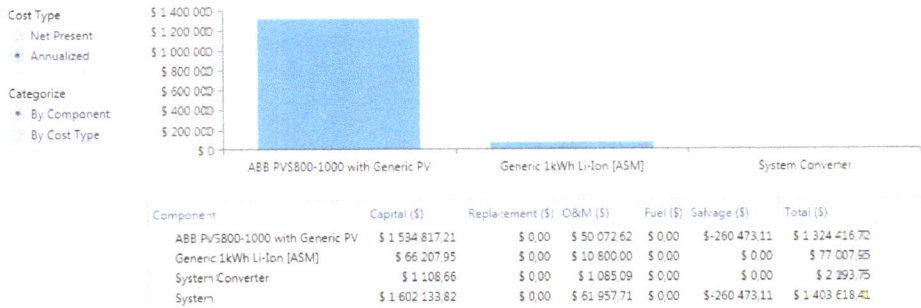

Figure 12. 5 MW PV plant annualised cost—batteries off-grid.

4.2. Case Study 2: 5 MW Load, PV Plant and 20% Pump Storage Plant

The PV plant and the load used in the previous section were decreased to match the 5 MW load. Storage batteries are very flexible in terms of where they can be used (different locations). Any location may be suitable, unlike the location for pumped storage where it is restricted to areas where there is a huge amount of water and high altitudes. For this study, we assumed that the optimum location for Pump storage was Palmiet, in the Western Cape since there is already similar storage existing in the area. A schematic diagram of PV plant with PPS as base storage is shown on Figure 13 below.

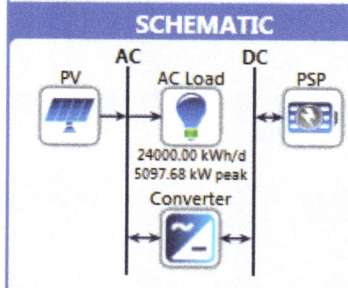

Figure 13. Pumped storage (PPS) vs. 5 MW PV plant schematic.

D_{cap} is the Dam Capacity (upper and lower) to be calculated

T_d is the time required to discharge and restore water to upper dam = 12 h, since this water had to be available for electrical production the next day

H_{eff} is the effective head = 150 m

η_{gen} is the generator efficiency = 90%

g is the gravitational force = 9.81 m/s^2

Dis_{rate} is the discharge rate/flow = 0.03 cubic metres per s.

Since the load was already known, some of the above was assumed in order to calculate the amount of water or dam capacity to meet the load demand:

L_{peak} is the peak load = 5 MWh

PPS rating to be 20% of the load = 1 MWh.

It is assumed one generator output generation over 12 h will be 245 kW; this will be used only as a baseline. Bigger generators are normally used in real life situations. HOMER has its pumped storage per generator stored at 24 kWh, this helped on the calculations below in order to use the program effectively;

Therefore $\frac{1 \text{ MWh}}{245 \text{ kWh}}$ = 4 generators, was practical. It was possible to use one generator of 250 kW or higher ratings to reduce the number to a maximum of 4 units.

Discharging;

To discharge 1 MW in 12 h is required:

$$\text{Power generated} \left(P_{gen}\right) = \frac{Energy}{Time} = 1 \text{ MWh}/12 \text{ h} = 83.33 \text{ kW} \tag{18}$$

$$\text{Mass of water required} = D_{cap} = \frac{P_{gen}}{g \times H_{eff} \times Dis_{rate} \times \eta_{gen}} \tag{19}$$

$D_{cap} = \frac{83.33 \times 10^3}{9.81 \times 150 \times 0.03 \times 0.9} = 2097$ m^3 of water required.

Charging;

Since the same turbine could be used as a pump, the effective head, efficiency and power remained unchanged. Therefore, the flow *rate* was calculated as follows:

$$\text{Flow rate} = F_{rate} = \frac{P_{gen} \times \eta_{gen}}{g \times H_{eff}} \tag{20}$$

$F_{rate} = \frac{83.33 \times 0.9}{9.81 \times 150} = 0.06$ m^3 per s.

$$\text{Time required to refill the upper dam} = T_{refill} = \frac{D_{cap}}{F_{rate} \times time} \quad (21)$$

$T_{refill} = \frac{2097}{0.06 \times 3600} = 9.7$ h.

$$\text{Electrical energy required} = (P_{gen}) \times (T_{refill}) \quad (22)$$

83.33 kW × 9.7 h = 0.81 MWh.

Round trip efficiency of the PPS = ratio of discharging electrical energy output to the charging electrical input = 1/0.81 = 1.23

$$\text{Maximum Capacity} = \frac{Energy}{Voltage} = \frac{1\text{MWh}}{240\text{V}} = 4.167 \times 10^3 \text{ Amp h} \quad (23)$$

Using pumped storage to supplement the load and rectify the PV seems to be working; the cost associated with the system of only 20% storage capacity was just USD 14 million as illustrated in Figures 14 and 15. Storage costs were quite high on PPS compared to the batteries. Battery storage cost is USD 13 million while PPS is USD 14 million. Based on cost alone, for a 5 MW plant with 20% storage, batteries were considered the better option. The same was true for annualised costs, thus PPS is quite expensive storage.

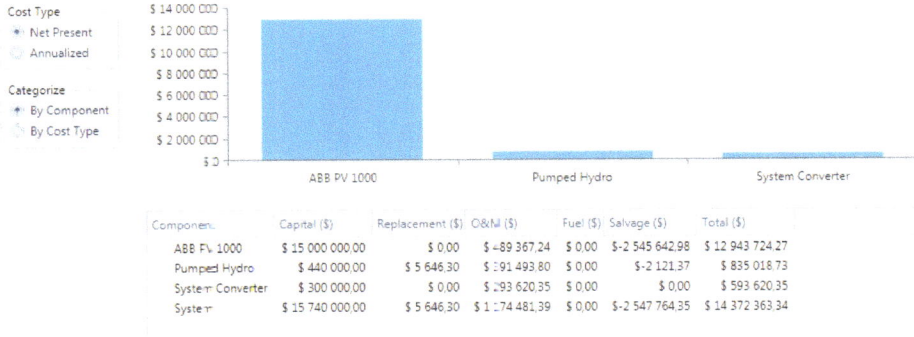

Figure 14. 5 MW PV plant net cost analysis—pumped storage plant.

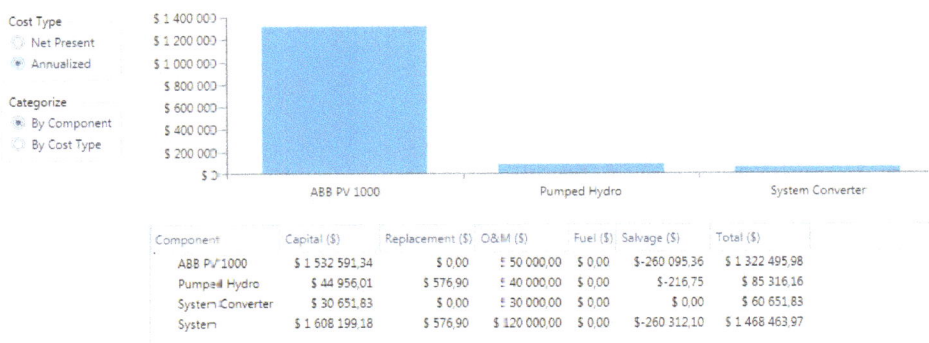

Figure 15. 5 MW PV plant annualised cost analysis—pumped storage plant.

Table 1. Storage cost comparison on 5 MW PV plant.

Cost Type	Storage Batteries System (Case Study 1)	PPS System (Case Study 2)	System Difference = (Battery − PPS)
Net Present Cost	USD 13,737,697.00	USD 14,372,363.34	USD −6,346,666.34
Annualised Cost	USD 1,403,618.41	USD 1,468,463.77	USD −64,845.3

Reading from Table 2 below, it can be seen that the difference between the two components is not big if analysed by individual components in the system. Meaning that on bigger plants the individual component does not play a big role but the entire system required to accommodate that individual component. Comparing Tables 1 and 2, it can be concluded that when building the system consider all components required to build it. If calculating only the cost of the component, then incorrect information will be used.

Figures 16 and 17 depict the operation of the plant/system: PV + PPS = load. Technically, the system was configured to work in the following manner:

PV is the primary source of power. From 6 am, the PV supplies the load with assistance/backup from the PPS, while at the same time, the PV is pumping water back to the upper reservoir. At 9 am the PV is at maximum output and the hydro is no longer supplying the load until the moment where there is surge on the load and the PV cannot cope with it. During the day from 12 pm, when the PV is stable, the excess electricity is used to pump water back to the upper reservoir. At that time no water is discharged until there is a sudden load surge on the system again. In the afternoon during peak time, the PV cannot cope with the usual peak load increase and, simultaneously, the solar energy decreases towards sunset. At this time the PPS takes over and supplies the load overnight while the PV is off. This continues until the following morning at 7 am when the PV picks up again.

Table 2. Storage cost comparison of individual components on 5 MW PV plant.

Cost Type	Storage Battery Bank	PPS Storage Bank	Component Difference = (Battery − PPS)
Net Present Cost	USD 753,703.22	USD 835,018.73	USD −81,315.51
Annualised Cost	USD 77,007.95	USD 85,316.16	USD −8,308.71

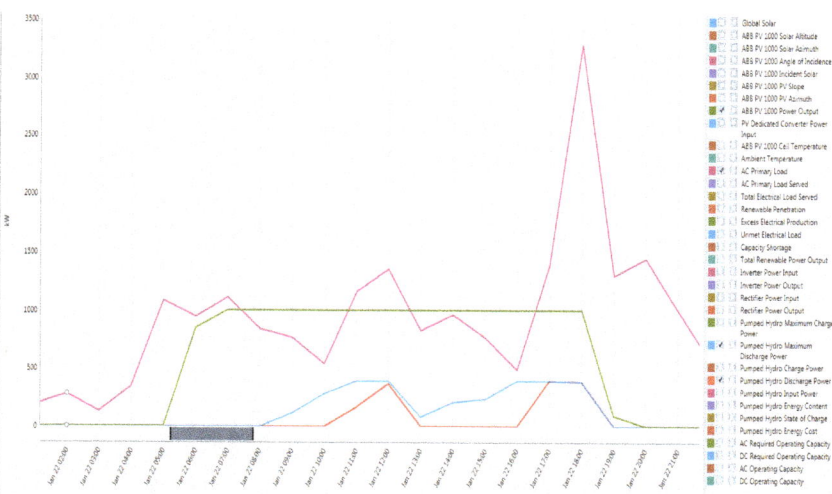

Figure 16. Hourly performance of PPS on 5 MW PV plant.

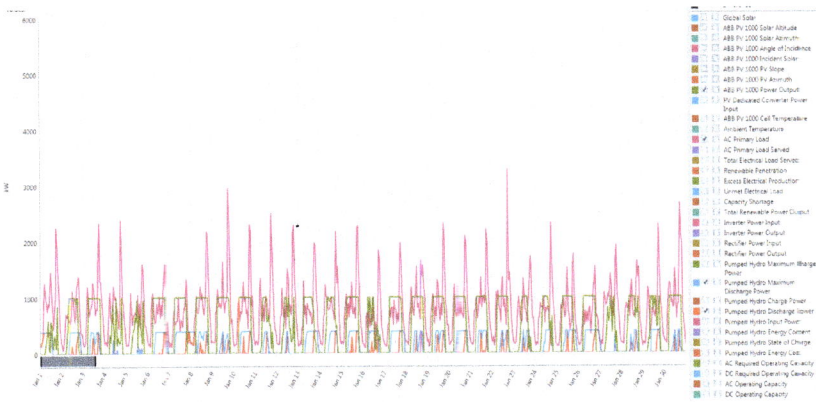

Figure 17. PPS monthly technical analysis on 5 MW PV plant.

4.3. Case Study 3: 3 MW Load, PV Plant and 20% Storage Battery Capacity

The peak load was 3 MWh, while just 20% of the load, which is only 600 kWh, had to be stored on Li-Ion batteries. A total of 600 batteries would be required to make 600 kWh storage. The characteristics of the battery are still the same as Figure 5. The schematic diagram of PV plant and Li-ion battery bank is shown on Figure 18 below.

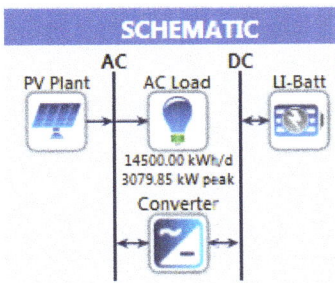

Figure 18. Case study—3 MW PV plant with 20% Li-ion bank, and load connected—off-grid.

As in the previous case study, the same results can be realised from a decreased system in terms of performance. Figures 19 and 20 above illustrate how the PV would supply the load throughout the year, with storage facility and peak months being July. It is clear that PV behaviour is not changed by loading or plant scaling; it will continue to supply the load at times where there is sun available and be off at night. Since the simulation was based in the Cape Town area, the sun is usually up at 5:34 am in the summer and sets around 20:00 as illustrated in Figure 20, which is usually from 5 am till 6 pm South African Time. At night there is no power supplied to the load. On other days, the output is not smooth throughout the day; this can be seen at 13:00 in Figure 20, this is the point at which storage becomes useful.

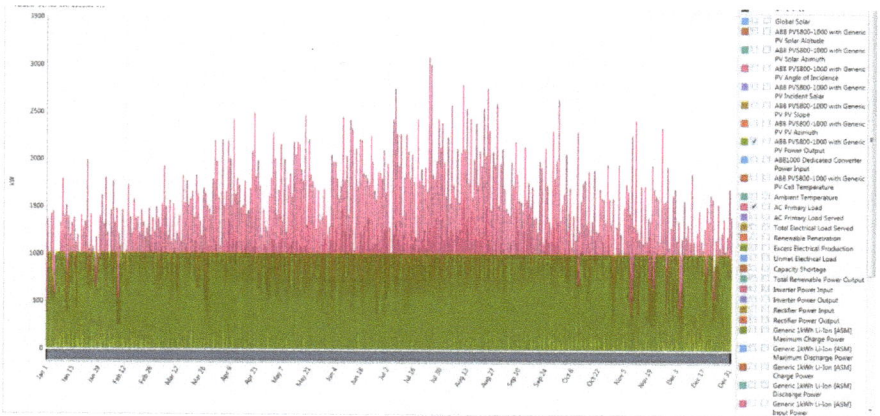

Figure 19. 3 MW PV plant output without storage yearly performance.

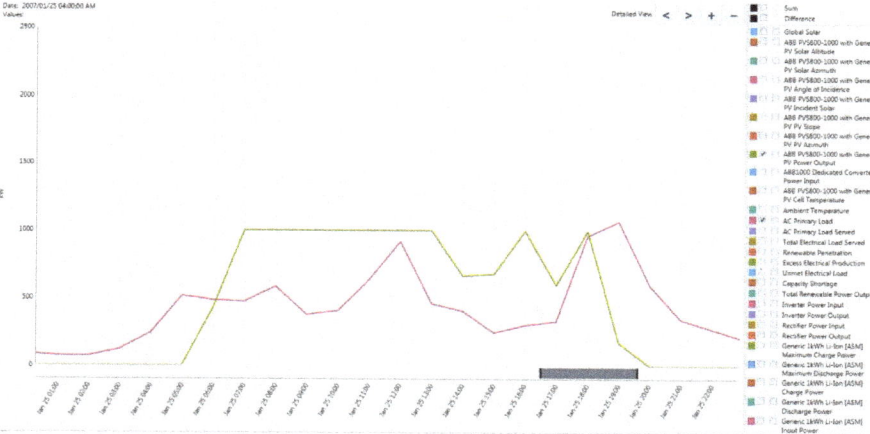

Figure 20. 3 MW PV plant output without storage day analysis.

As may be observed from Figure 21

- PV power output is drawn as a green line in the graph;
- The load is the purple line in the graph;
- Maximum discharge battery output is drawn as a blue line; and
- The operating discharge output is in red on the graph.

When analysing the two graphs below (i.e., Figures 21 and 22), it may be seen that the PV output started increasing from 5 am and peaks at 6 am with an output power of 1000 kW. The unmet and fluctuating load remained present for 24 h. Li-ion battery banks were available and could be discharged from 7 am. if required, as indicated by the blue line of the graph, but this was not the case because, at the time, PV was peak and could manage the load. The graph illustrates that the maximum available storage that could be discharged was 446 kW. The red line indicates the operating battery storage operation, where there was no discharge on batteries throughout the day, until such time where the load was greater than the PV output. That scenario occurs at 17:00 in the afternoon. At that time the peak load started, the battery 'kicked-in' to supplement the PV up until 18:00. At 18:00, PV was peak at 1000 kW; Load was 1983 kW, while the storage supply peak was at 446 kW. The storage battery percentage overload = 446/1983 = 22% peak. After 18:00, the load starts decreasing and the storage facility and the PV supply followed. By 19:00, both the battery bank and the PV plant stop supplying the load as required. The scenario is repeated daily throughout the year.

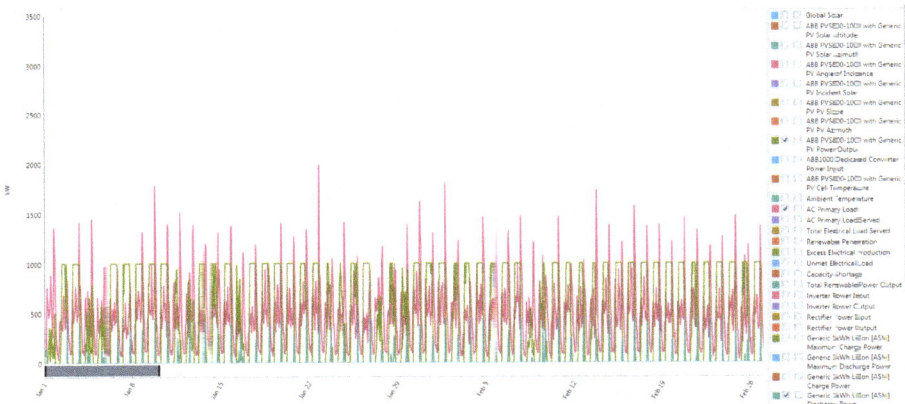

Figure 21. 3 MW PV plant output with storage battery month preview.

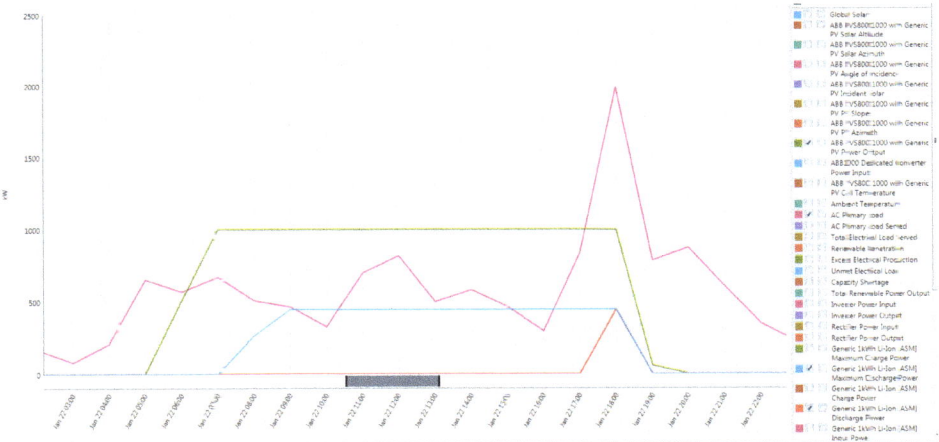

Figure 22. 3 MW PV plant and Li-ion output on summer day.

4.4. PV-Battery Plant Off-Grid Cost Analysis

Based on the results of above Figures 23 and 24, it can be seen that having a 3 MW PV power plant with 20% battery storage capacity in the Western Cape would cost approximately Net USD 8 million with an annual cost of USD 0.8 million for a project with a life cycle of 25 years. This system is technically excellent but could require an exorbitant amount of funding to build. This system is 100% renewable with 20% battery storage capacity, 48.3% unmet load and up to 73% capacity shortage. Nonetheless, the return on investment would be very low at −0.8%, possibly making this too expensive to construct. However, it can be used as a baseline for smaller systems/plants. Since the system was off-grid, the quantity of batteries has increased. Accordingly, it is proven that there is no investment reward for building such a system as it has very low investment rate.

The difference between using a system with no storage on the same system is only USD 540,000.

Component	Capital ($)	Replacement ($)	O&M ($)	Fuel ($)	Salvage ($)	Total ($)
ABB PVS800-1000 with Generic PV	$ 9 000 000,00	$ 0,00	$ 293 620,35	$ 0,00	$-1 527 385,79	$ 7 766 234,56
Generic 1kWh Li-Ion [ASM]	$ 360 000,00	$ 223 210,21	$ 58 724,07	$ 0,00	$-31 508,24	$ 610 426,04
System Converter	$ 180 000,00	$ 0,00	$ 176 172,21	$ 0,00	$ 0,00	$ 356 172,21
System	$ 9 540 000,00	$ 223 210,21	$ 528 516,62	$ 0,00	$-1 558 894,02	$ 8 732 832,81

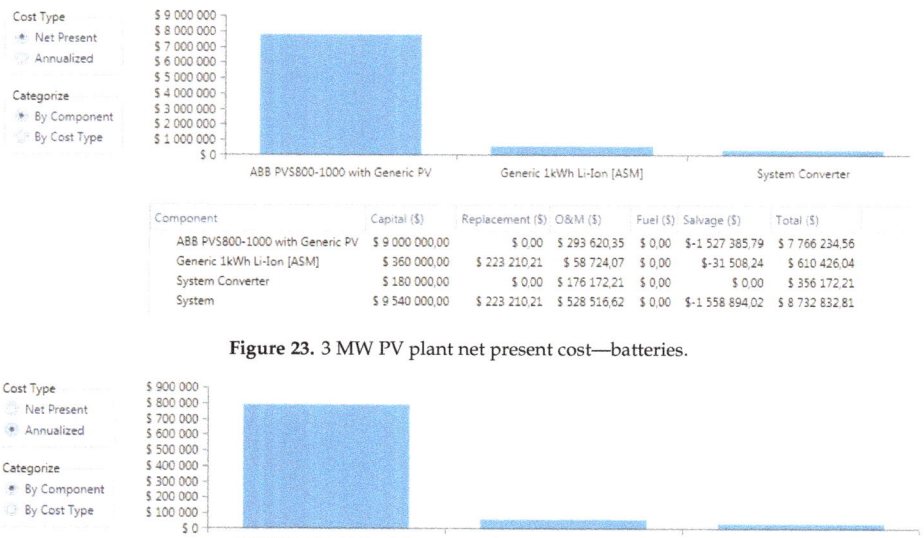

Figure 23. 3 MW PV plant net present cost—batteries.

Component	Capital ($)	Replacement ($)	O&M ($)	Fuel ($)	Salvage ($)	Total ($)
ABB PVS800-1000 with Generic PV	$ 919 554,80	$ 0,00	$ 30 000,00	$ 0,00	$-156 057,21	$ 793 497,59
Generic 1kWh Li-Ion [ASM]	$ 36 782,19	$ 22 806,00	$ 6 000,00	$ 0,00	$-3 219,28	$ 62 368,91
System Converter	$ 18 391,10	$ 0,00	$ 18 000,00	$ 0,00	$ 0,00	$ 36 391,10
System	$ 974 728,09	$ 22 806,00	$ 54 000,00	$ 0,00	$-159 276,50	$ 892 257,59

Figure 24. 3 MW PV plant annualised cost—batteries off-grid.

4.5. Case Study 4: 3 MW Load, PV Plant and 20% Pump Storage Plant

The PV plant and the load used in case study 2 were decreased to match the 3 MW load. As per case study 2, the optimum location of the system remains the same. A schematic diagram of 3 MW load supplied by PV plant and PPS storage is shown in Figure 25 below;

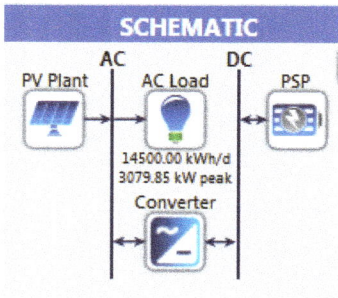

Figure 25. Pumped storage (PPS) vs. 3 MW PV plant.

D_{cap} is the Dam Capacity (upper and lower) to be calculated.
T_d is the time required to discharge and restore water to upper dam, which is 12 h, since this water had to be available for electrical production the next day.
H_{eff} is the effective head = 150 m
η_{gen} is the Generator efficiency = 90%
g is the gravitational force = 9.81 m/s^2

Dis_{rate} is the discharge rate/flow = 0.03 cubic metres per s.

Since the load was already known, some of the above were assumed to be able to calculate the amount of water or dam capacity to meet the load demand.

L_{peak} is the peak load = 3 MWh
PPS rating to be 20% of the load = 0.6 MWh.

Assuming one generator output generation over 12 h to be 245 kW; this will be used just as a baseline. Bigger generators are normally used in real life situations.

Therefore $\frac{0.6 \text{ MWh}}{245 \text{ kWh}}$ = 2.44 generators, which is impractical. One generator of 300 kW or higher ratings can be used to reduce the number to a maximum of 2 units.

Discharging

To discharge 600 kW in 12 h is required; using Equation (18); Power generated (P_{gen}) = 600 kWh/12 h = 50 kW

Using Equation (19); mass of water required = $D_{cap} = \frac{P_{gen}}{g \times H_{eff} \times Dis_{rate} \times \eta_{gen}} = \frac{50 \times 10^3}{9.81 \times 150 \times 0.03 \times 0.9}$ = 1258 m³ water is required.

Charging

Since the same turbine could be used as a pump, as discussed in Chapter 3, the effective head, efficiency and power remained unchanged. Therefore, the flow rate was calculated as follows:

From Equation (20); flow rate = $F_{rate} = \frac{P_{gen} \times \eta_{gen}}{g \times H_{eff}}$

$F_{rate} = \frac{50 \times 0.9}{9.81 \times 150}$ = 0.031 m³ per s

Time required to refill the upper dam, using Equation (21) = $T_{refill} = \frac{1258}{0.06 \times 3600}$ = 5.82 h

From Equation (22); Electrical energy required = 50 kW × 5.82 h = 291 kWh

Round trip efficiency of the PPS = ratio of discharging electrical energy output to the charging electrical input = 0.6/0.291 = 2.06

Maximum capacity using Equation (23) = $\frac{0.6 \text{ MWh}}{240 \text{ V}}$ = 2500 Amp hours. This is the maximum electrical output divided by the nominal voltage.

Using pumped storage to supplement the load and rectify the PV seemed to be working; the cost associated with the system at just 20% storage capacity was just USD 8.6 million as recorded in Figures 26 and 27. Storage costs are almost the same for the PPS as that of the batteries in Figures 23 and 24. The difference between the two systems presented in Table 3 below. Based on the results, it can be stated that any storage facility can be chosen for the 3 MW plant.

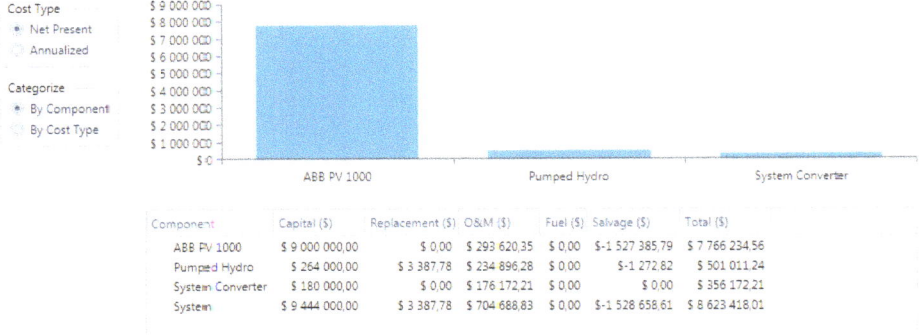

Figure 26. 3 MW PV plant net cost analysis—pumped storage plant.

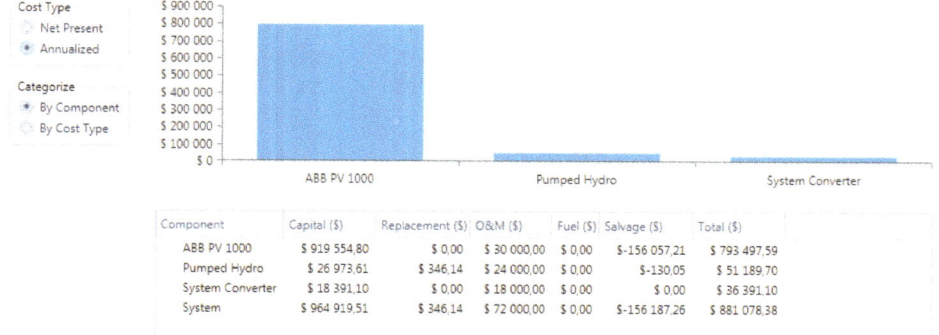

Figure 27. 3 MW PV plant annualised cost analysis—pumped storage plant.

Table 3. Storage cost comparison on 3 MW PV plant.

Cost Type	Storage Batteries (Case Study 3)	PPS (Case Study 4)	Difference = (Battery − PPS)
Net Present Cost	USD 8,732,832.81	USD 8,623,418.01	USD 109,414.80
Annualised Cost	USD 892,257.59	USD 881,078.38	USD 11,179.21

Reading from Table 4 below, it can be seen that the difference between the two components is reduced when analysing by individual component in the system. Meaning that on bigger plants the individual component does not play a big role but the entire system required to accommodate that individual component. However, when the system reduced then the individual component makes a difference as seen in Tables 3 and 4.

Table 4. Storage cost comparison of individual components on 3 MW PV plant.

Cost Type	Storage Battery Bank	PPS Storage Bank	Component Difference = (Battery − PPS)
Net Present Cost	USD 610,426.04	USD 501,011.24	USD 109,414.80
Annualised Cost	USD 62,368.91	USD 51,189.70	USD 11,196.4

Figures 28–30 represent the operation of the plant/system with the PPS connection. PV + PPS = load. Technically, the system was configured to work in the following manner:

PV is the primary source of power. From 5 am. in the morning, the PV supplies the load with assistance/backup from the PPS just at peak times. During the night the PV storage discharges to supply the load or can be switched off. While supplying the load from 6 am, excess electricity is used to pump water back to the upper reservoir as per the calculation above. In the afternoon during peak time, the PV is unable to cope with the usual peak load increase and simultaneously, the solar energy decreases towards sunset. At this time the PPS takes over and supplies the load overnight while the PV is off. This continues until the following morning at 6 am. when the PV picks up and becomes stable. The operation remains the same as in case study 2.

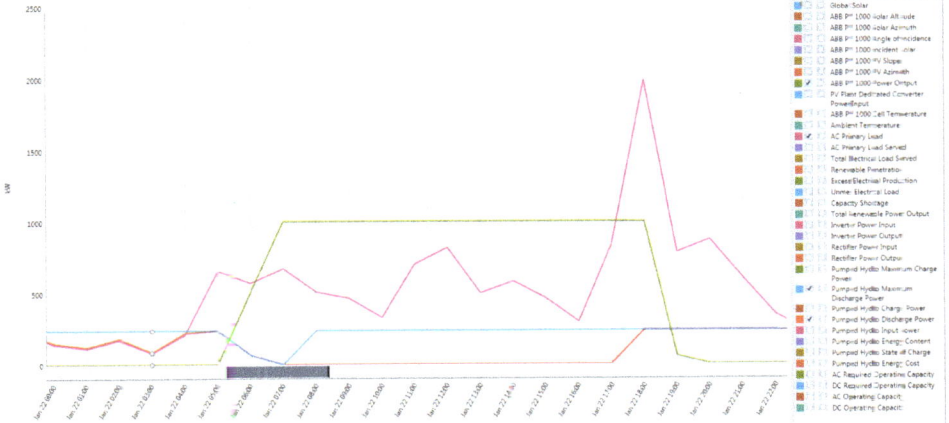

Figure 28. Hourly performance of PPS on 3 MW PV plant.

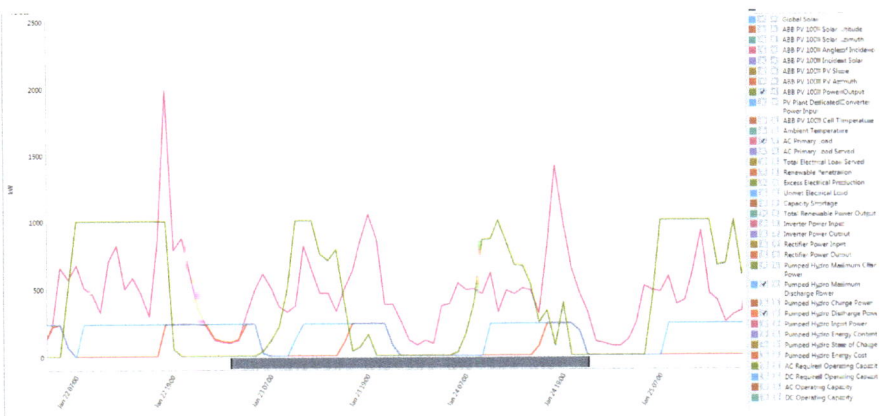

Figure 29. Daily performance of PPS on 3 MW PV plant.

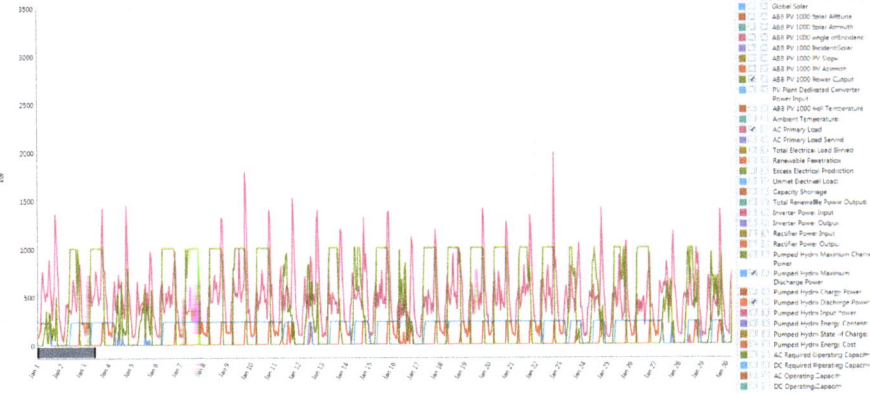

Figure 30. PPS monthly technical analysis on 3 MW PV plant.

4.6. Case Study 5: 1 MW Load, PV Plant and 20% Storage Battery Capacity

The peak load is 1 MWh, while just 20% of the load, which is only 200 kWh had to be stored on Li-ion batteries. The characteristics of the battery remain the same in this study as shown on Figure 5. A total of 200 batteries is required to make 200 kWh storage, the schematic diagram is shown in Figure 31.

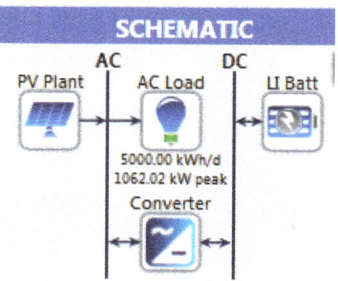

Figure 31. Case Study 5—1 MW PV plant, Li-ion bank, and load connected—off-grid.

The system is rather smaller than the two previous case studies; PV output is rated at 500 kW with a maximum demand of 1000 kW. Figures 32 and 33 above illustrates how PV will supply the load throughout the year with a storage facility of 20%. Previous case studies indicated the beginning and middle of the year, which are summer and winter. Case studies 5 and 6 will focus on springtime which is in September. PV plant behaviour is not changed by the loading or plant scaling; it will continue to supply the load at times when there is available sun and will switch off at night. Since the simulation was based on the Cape Town area, the sun in the springtime is usually up at 7 am and set at 19:00 as shown in Figure 32, that is usually from 5 am till 6 pm South African time. At night time there will be no power supplied to the load. On some days the output is not smooth throughout the day. As per Figure 33, there are two occasions where storage is required to supplement the PV: between 6 and 10 am as well as between 6 and 8 pm. In this scenario, PV would produce excess energy during the day, which may allow the charging of batteries. This means that batteries can be used more at night since there is enough power to charge during the day in this scenario.

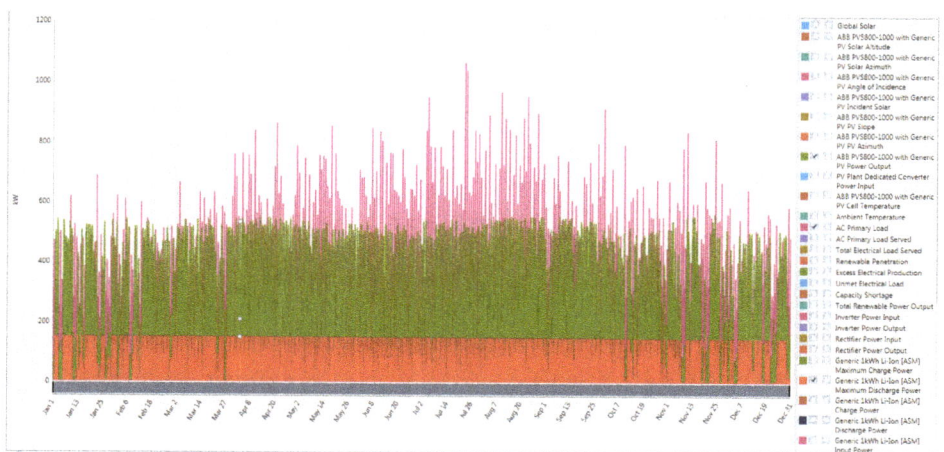

Figure 32. 1 MW PV plant output with battery storage yearly performance.

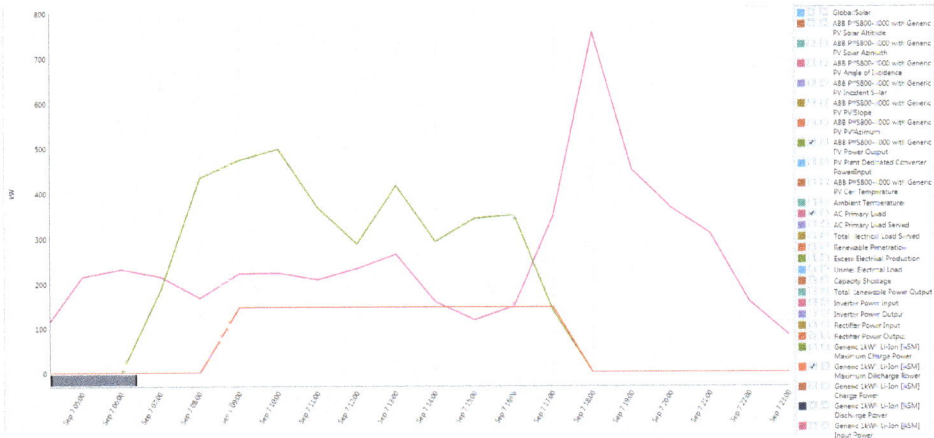

Figure 33. 1 MW PV plant output with battery storage day analysis.

As seen in Figure 34:

- PV power output is drawn in a green line in the graph;
- The load is drawn in purple in the graph;
- Maximum discharge battery output is the red line in the graph; and
- The operating discharge output is the blue line in the graph.

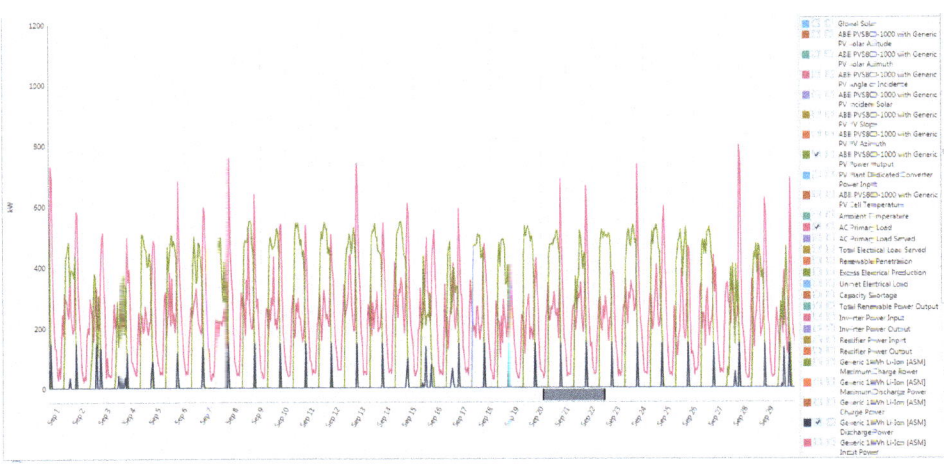

Figure 34. 1 MW PV plant output with storage battery month preview.

Analysing the two graphs below (i.e., Figures 34 and 35), it can be seen that PV output starts increasing from 5 am and peaks at 10 am with an output power of 460 kW. The unmet and fluctuating load has remained present for a few hours during the day and 11 h at night. Li-ion Battery banks are available and can be discharged from 9 am if required as shown by the red graph line, but this does not occur since at the time PV is at peak and can manage the load until 11 am. The battery line in the graph reveals that the maximum available storage that can be discharged is 145 kW. The blue graph line showing the operating battery storage operation, where there will be no discharge from batteries throughout the day until such time where the load is greater than PV output. That scenario is seen between 11:00 and 13:00 in the morning to the afternoon. Then again between 15:00 and 17:00. After 17:00, peak load starts, battery power is already utilised during the day and can no longer be used, and PV output is also declining since it is sunset. Looking into peak load at 12:00, PV is peak 358 kW; load is 423 kW while the storage supply is peak 69 kW. Storage battery percentage overload = 69/423 = 16% peak. The scenario will be repeated daily throughout the year with different seasons as per with respective to load profile graphs in Figure 36 and battery discharge power in Figure 37 below.

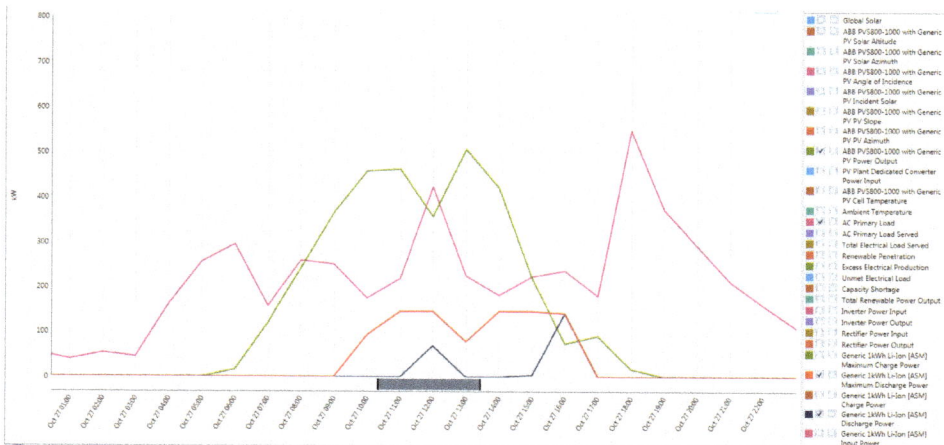

Figure 35. 1 MW PV plant and Li-ion output on October spring day.

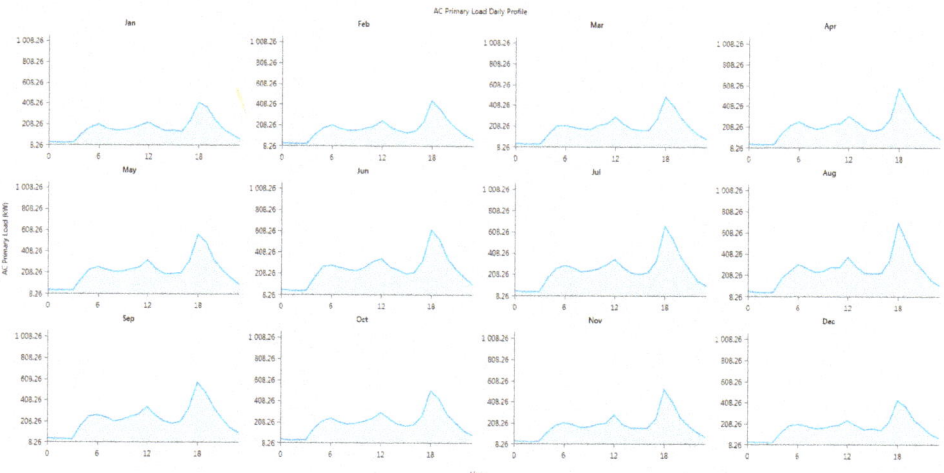

Figure 36. 1 MW annual load profile.

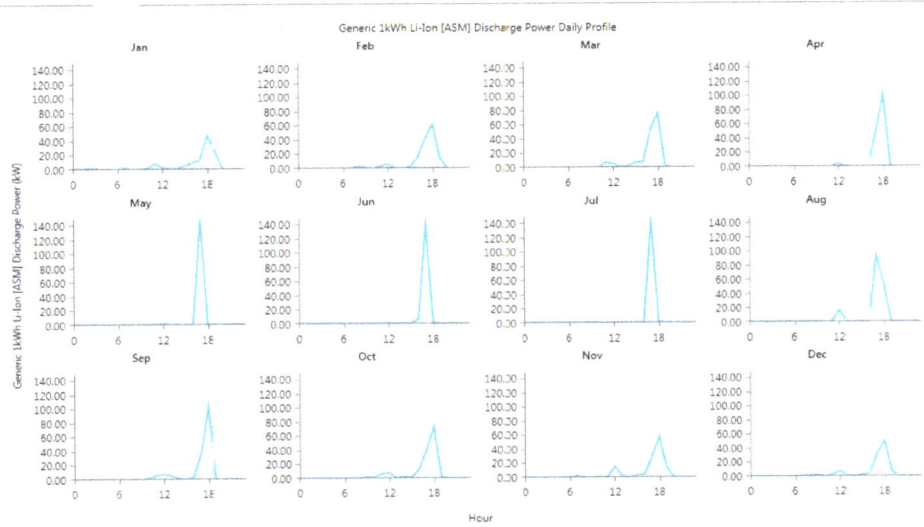

Figure 37. 200 kWh battery discharge power per hour profile for a year.

4.7. PV-Battery Plant Off-Grid Cost Analysis

Based on the below results in Figures 38 and 39, it can be seen that having a 1 MW PV power plant with 20% battery storage capacity in the Western Cape will cost approximate Net USD 1.6 million with annual cost of USD 163,000 for a project with the life cycle of 25 years. This system is technically excellent and may require little capital to build it. This system is 100% renewable with 20% battery storage capacity. However, the return on investment is very low at −0.7%. This then may be too expensive to construct because of no return on investment. However, it can be used as a base for smaller systems/plants. Since the system is off-grid, the number of batteries can be reduced or increased as required.

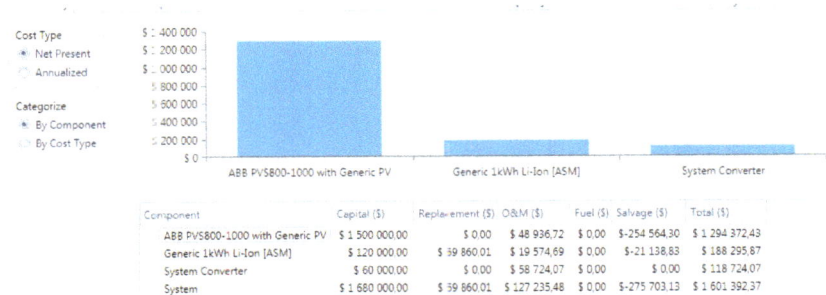

Figure 38. 1 MW PV plant net present cost PV—batteries.

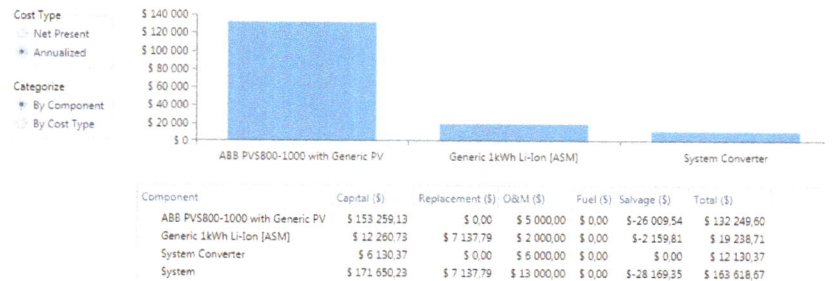

Figure 39. 1 MW PV plant annualised cost—batteries off-grid.

The difference between using a system with no storage on the same system is only USD 400,000.

4.8. Scenario 6: 1 MW Load, PV Plant and 20% Pump Storage Plant

The PV plant and the load used in the previous section of this chapter was decreased to match the 1 MW load. Storage batteries are very flexible in terms of where they can be used (different locations). Any location will work, unlike the pumped storage where it is restricted to areas where there is a huge volume of water and high altitudes. For this study, as mentioned, we assumed the optimal location for pump storage is Palmiet in the Western Cape. Figure 40 shows schematic diagram of scenario 6.

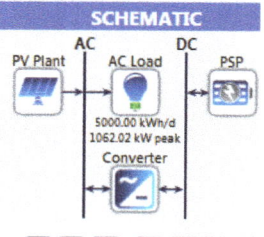

Figure 40. Pumped storage (PPS) vs. 1 MW PV plant.

D_{cap} is the Dam Capacity (upper and lower) to be calculated.
T_d is the time required to discharge and restore water to upper dam = 12 h, since this water had to be available for electrical production the next day.
H_{eff} is the effective head = 150 m
η_{gen} is the Generator efficiency = 90%
g is the gravitational force = 9.81 m/s^2
Dis_{rate} is the Discharge rate/flow = 0.03 cubic metres per s.

Since load was already known, some of the above were assumed to be able to calculate the amount of water or dam capacity to meet the load demand.

L_{peak} is the Peak load = 1 MWh
PPS rating to be 20% of the load, it is 200 kWh.

Assuming one generator output generation over 12 h to be 245 kW; this will be used only as a baseline. Bigger generators are normally used in real life situations.

Therefore $\frac{200 \text{ kWh}}{245 \text{ kWh}}$ = 1 generator which was enough. One generator of between 200–300 kW could be used or, for maintenance purposes, two generators of 150 kW could be utilised so that total power is not lost, should faults occur or maintenance be required.

Discharging

From Equation (18): to discharge 200 kW in 12 h; Power generated (P_{gen}) = 200 kWh/12 h = 17 kW

From Equation (19); mass of water required = $D_{cap} = \frac{P_{gen}}{g \times H_{eff} \times Dis_{rate} \times \eta_{gen}} = \frac{17 \times 10^3}{9.81 \times 150 \times 0.03 \times 0.9} = 427$ m^3 water is required.

Charging

Since the same turbine could be used as a pump, as discussed in Chapter 3, the effective head, efficiency and power remained unchanged. Therefore, the flow rate was calculated as follows:

Using Equation (20); flow rate = $F_{rate} = \frac{P_{gen} \times \eta_{gen}}{g \times H_{eff}} = \frac{17 \times 0.9}{9.81 \times 150} = 0.0104$ m^3 per s.

Using Equation (21); time required to refill the upper dam = $T_{refill} = \frac{427}{0.06 \times 3600} = 2$ h

Electrical energy required using Equation (22) = 17 kW × 2 h = 34 kWh

Round trip efficiency of the PPS = ratio of discharging electrical energy output to the charging electrical input = 0.2/0.34 = 0.59

Maximum Capacity from (23) = $\frac{200 \text{ kWh}}{240 \text{ V}} = 833$ Amp h.

Using pumped storage to supplement the load and rectify the PV also worked as did battery storage; the cost associated with the system at only 20% of storage capacity is just USD 2.7 million as shown in Figure 41. PPS storage cost is higher than when using the batteries on the same system as displayed in Figures 41 and 42. The difference between the two systems is reported in Table 5 below. Based on the cost results, the sole system that can be chosen for the 1 MW plan is storage batteries.

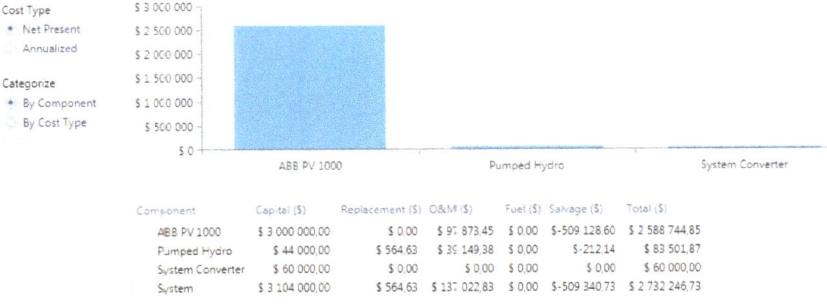

Figure 41. 1 MW PV plant net cost analysis—pumped storage plant.

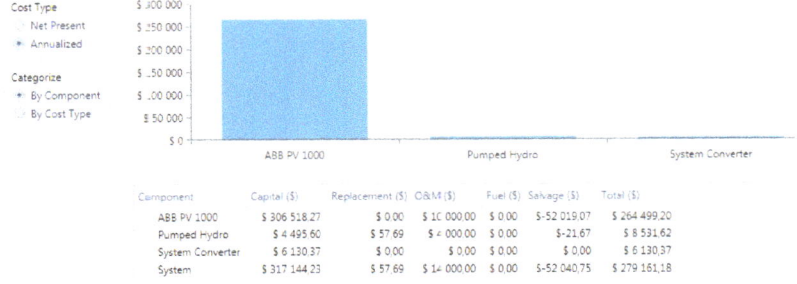

Figure 42. 1 MW PV plant annualised cost analysis—pumped storage plant.

Table 5. Storage cost comparison on 1 MW PV plant.

Cost Type	Storage Batteries (Case Study 5)	PPS (Case Study 6)	Difference = (Battery − PPS)
Net Present Cost	USD 1,601,392.37	USD 2,732,246.73	USD −1,130,854.36
Annualised Cost	USD 163,618.67	USD 279,161.18	USD −115,542.51

Reading from Table 6 below, it can be seen that the difference between the two components has reduced by reducing the storage capacity. However, the component cost is almost equal to the difference when comparing with case studies 3 and 4. This means that as the storage capacity reduces, the difference in component price also reduces. Though it is much cheaper to build the PPS system than storage batteries as shown on both Tables 5 and 6.

Figures 43 and 44 illustrate the performance of PPS on the 1 MW PV plant. As indicated previously, pumped storage works in such a way that during peak times and night hours, energy can be discharged to supply the load. During the day when the PV is overproducing power, the excess energy can be utilised to pump water back to the upper reservoir. The two figures above demonstrate the same can be used when the storage batteries are in use. In this situation, there is enough energy produced during the day, as shown in Figure 43, to be used as required. The same excess power can be sold to the grid to make profit and returns on the PV plant.

Figures 45–48 These figures indicate more charge and discharge power of PPS throughout the year. It can be seen that during the month of May, June and July the rainfall in the country is less and it is thus difficult to depend on PPS up until August when it starts raining again. Even though pumped storage relies on stored water, this water is subject to evaporation and can be lost in other forms.

Figure 48 shows the loading per hour on the PV plant and the peak times at which PPS will be required. It can be seen that different types of loading are experienced for each month and peak times change according to the month.

Table 6. Storage cost comparison of individual components on 1 MW PV plant.

Cost Type	Storage Battery Bank	PPS Storage Bank	Component Difference = (Battery − PPS)
Net Present Cost	USD 188,295.87	USD 83,501.87	USD 104,794.00
Annualized Cost	USD 19,238.71	USD 8,531.62	USD 10,707.09

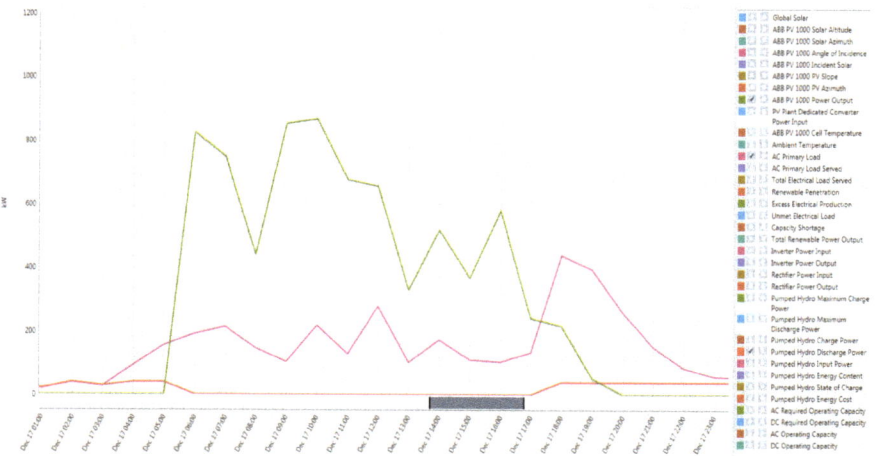

Figure 43. Hourly performance of PPS on 1 MW PV plant.

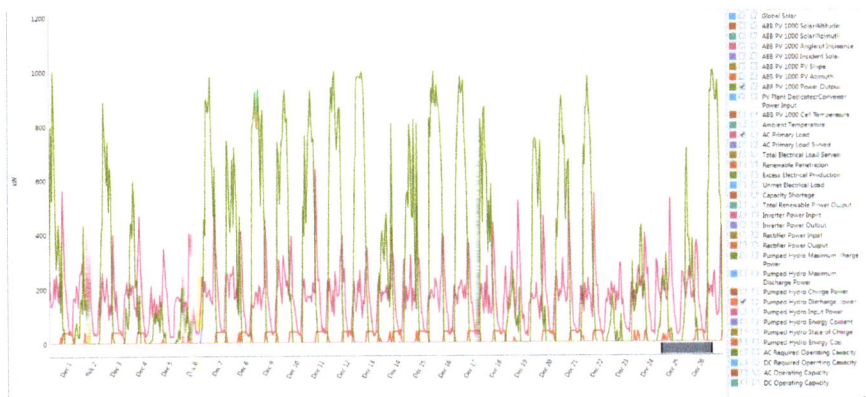

Figure 44. Daily performance of PPS on 1 MW PV system.

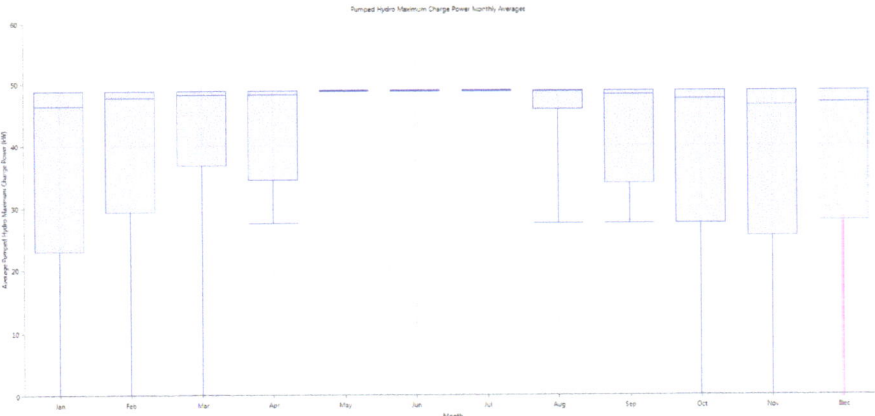

Figure 45. 200 kW PPS monthly maximum charge power.

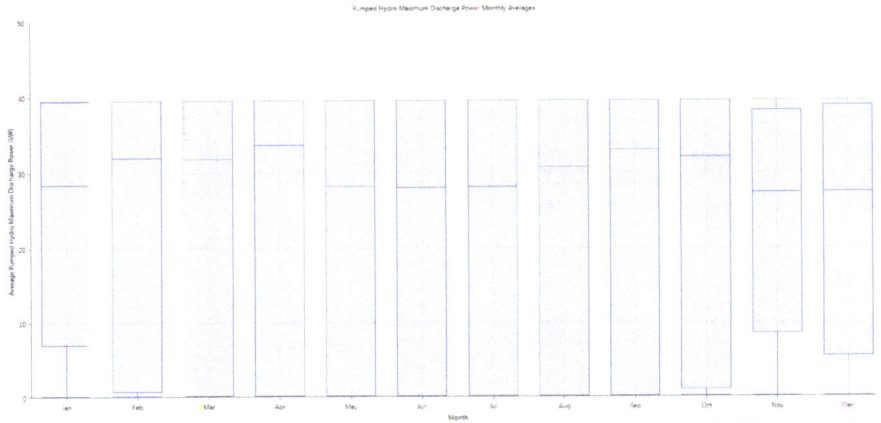

Figure 46. 200 kW PPS monthly maximum discharge power.

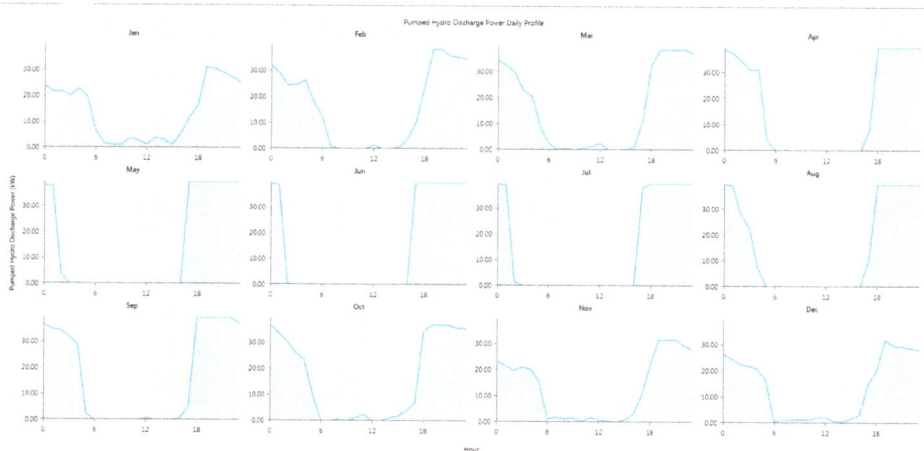

Figure 47. 200 kW PPS hourly discharge power profile.

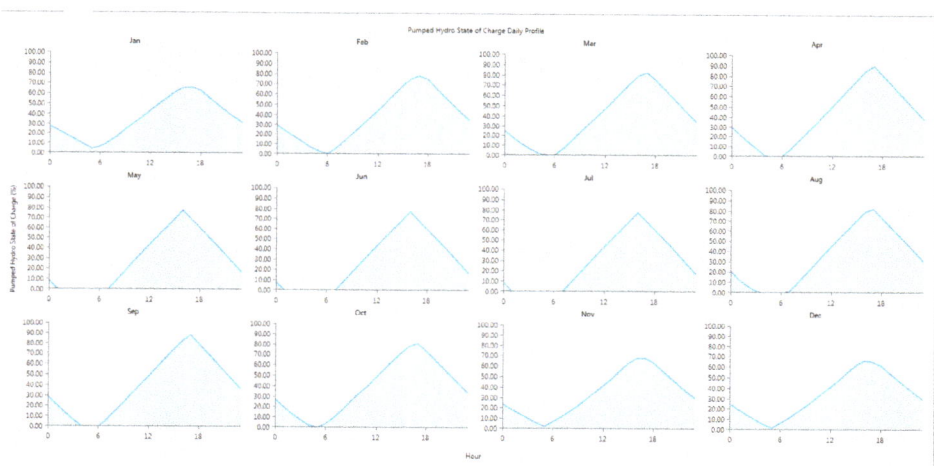

Figure 48. 200 kW PPS hourly state of charge profile.

Table 7 and Figure 49 below summarise the cost of storage for different power ratings; these exclude the PV plant and other components connected to the plant, such as converters. It can be seen that pumped storage is more expensive as the storage capacity increase and the battery becomes cheaper. For lower storage requirements, pumped storage is suitable and for higher load requirements batteries are the preferred or better option, cost-wise.

Table 7. Storage cost summary.

Scenario Type	Net Cost	Annualised Cost
1 MWh Lithium Ion Batteries	USD 753,703.32	USD 77,007.95
1 MWh Pumped Storage	USD 835,018.73	USD 85,316.16
600 kWh Lithium Ion Batteries	USD 610,426.04	USD 62,368.91
600 kWh Pumped Storage	USD 501,011.24	USD 51,189.70
200 kWh Lithium Ion Batteries	USD 188,295.87	USD 19,238.71
200 kWh Pumped Storage	USD 83,501.87	USD 8,531.62

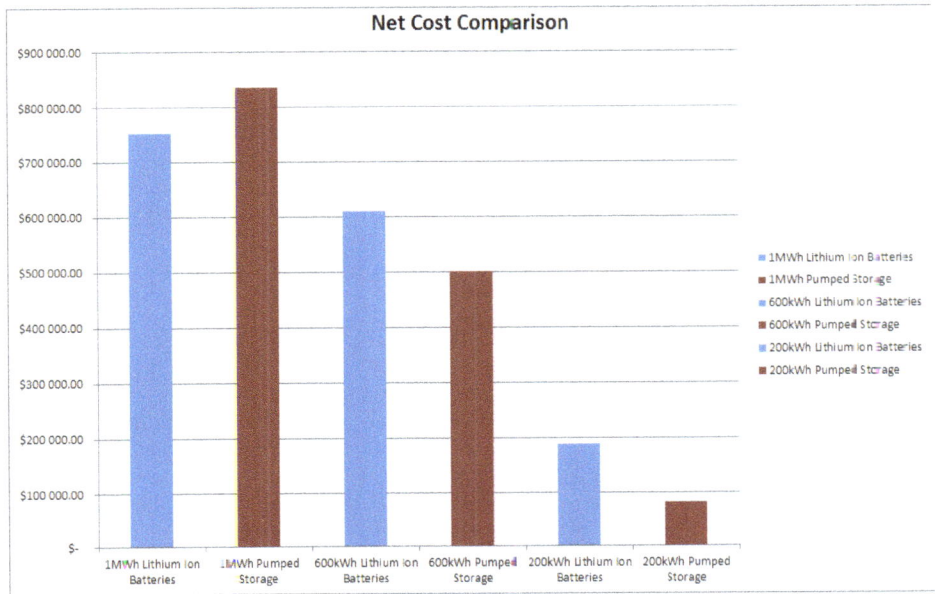

Figure 49. Net cost summary and comparison of storage plants.

5. Conclusions

To achieve the objectives of the study which include finding the economic and technical solution to renewable energy storage, the following design process was implemented—development of a PV model and its integration with storage batteries using lithium-ion; a pumped storage model was designed as well as its integration with the PV model. Comparison of the battery and PPS model was carried out and finally a developed model arising from the results obtained at different stages of design implementation and efficacy was realised. The model development and simulations were carried out in HomerPro® environments, Version 64x3.10.0 (Pro Edition). Analysis of the results were also carried to accentuate the impact and contribution of the various scenarios, factors and storage systems. Storage batteries and PPS were simulated and compared using 3 scenarios, which are 200 kWh, 600 kWh and 1 MWh storage capacities.

From the results obtained, it can be inferred that both storages are significant to PV output stabilisation as required. They both can be used to supplement PV in times of high demand and times where PV is not sufficient due to weather conditions or fluctuations. These storages have capacities to discharge anytime when required, although PPS has very rapid response and can be used where there are large loads or emergency loading is required. When operated in a hybrid system with PV, it was proven that for lower storage requirements such as 200 kWh storage bank, PPS is a cheaper option compared to storage batteries by almost half the cost. As storage capacity increases to 600 kWh then the gap is closing down and PPS is approximately 16% cheaper than batteries. The difference becomes lesser as storage capacity increase to 1 MWh.

The research has proven that when storage costs go higher than 1 MWh, then batteries can be selected. The assumption was made that both storage resources can be available at the same place.

The research gives the utility the option to select which storage to apply based on the resource at the area where PV is installed. This will improve and stabilise the output demand, increase generation capacity, and consumers will have access to power most of the time that in turn reduces the electricity usage cost. It contributes to socio-economic development in the country, by creating jobs. Companies will have more usage period for lighting activities and that generally increases production capacity.

The solution is environmentally friendly and alleviates the present crisis of load shedding due to the unbalance and high demand as well as lower generations.

The research and technology trends indicated that lithium-ion is the future of energy storage. Pumped storage had a lower price per megawatt compared to batteries in 2016. Pumped storage costs ranged between $200–260 /MWh while batteries were $350–1000 /MWh. But the cost of batteries from 2016 to 2017 has also reduced by almost 25%. With the continued cost decline in lithium-ion batteries, pumped storage will be history. Reports indicate lithium-ion batteries may go lower to $120 /MWh in 2025. Taking into consideration equal long-life span (50–100 years) for pumped storage, battery costs may range between $200–300 /MWh.

The other point to consider is that cost comparison should be made not only on installed stored capacity. Most researchers tend to compare storage per MW using installed capacity and that is not necessarily usable capacity. The research was successful and proved that both pumped storage and batteries can be used depending on site location, it further proved that in future more batteries will be used and the size will continue to reduce making the transport cost lower. Also considering construction costs, batteries can be installed quicker than building a pumped storage plant.

Finally, it is proven that these storage facilities, i.e., lithium-ion batteries and pumped storage plants will stabilise the output demand; increase generation capacity; provide backup power; provide consumers with power for longer periods; and thus increase their production capacity and reduce the electricity cost since they will be using power from stable renewable energies instead of fossil fuels. The latter automatically contributes to socio-economic development in the country by creating jobs and saving the existing ones. Taking into consideration the construction and maintenance of the storage facility, some jobs will be created. Reducing the use of coal-fired stations will assist in the reduction of existing carbon emissions. The solution is environmentally friendly and alleviates the present crisis of load shedding due to the imbalance of high demand to lower generations. It is expected that other manufacturers will invest in cost-effective batteries instead of pumped storage as it has no real developmental future as water cannot be developed. With ever more strict rules regarding the care of the environment, it is expected that water usage can become restricted since every living species requires water.

Author Contributions: Conceptualization, S.P.D.C. and M.N.; Methodology, M.N. and S.P.D.C.; Software, M.N.; Validation, S.P.D.C. and M.N.; Formal Analysis, M.N. and S.P.D.C.; Investigation, M.N.; Resources, M.N.; Data Curation, M.N.; Writing—Original Draft Preparation, M.N.; Writing—Review and Editing, M.N., O.P., S.P.D.C.; Visualization, S.P.D.C., M.N., O.P.; Supervision S.P.D.C. and O.P.; Project Administration, M.N., S.P.D.C., O.P.; Funding Acquisition, O.P. and S.P.D.C.

Funding: This research was internally funded by the Tshwane University of Technology, Centre for Energy and Electric Power, Auto-X Pty Ltd. and F'SATI.

Acknowledgments: The authors would like to thankfully acknowledge the Tshwane University of Technology, Centre for Energy and Electric Power, Auto-X Pty Ltd. and F'SATI, Pretoria, South Africa for providing the research infrastructure for conducting this research.

Conflicts of Interest: The authors declare no conflict of interest.

References

1. PPIAF. *South Africa's Renewable Energy IPP Procurement Program: Success Factors and Lessons*; PPIAF: Washington, DC, USA, May 2014; Available online: http://www.gsb.uct.ac.za/files/ppiafreport.pdf (accessed on 25 July 2016).
2. Harque, A.; Rahman, M.A. Study of a solar PV-powered mini-grid pumped hydroelectric storage & its comparison with battery storage. In Proceedings of the 2012 7th International Conference Electrical & Computer Engineering (ICECE), Dhaka, Bangladesh, 20–22 December 2012; pp. 626–629.
3. Chen, G.Z.; Liu, D.Y.; Wang, F.; Ou, C.Q. Determination of installed capacity of pumped storage stations in WSP hybrid power supply system. In Proceedings of the Sustainable Power Generation and Supply (SUPERGEN'09), International Conference, Nanjing, China, 6–7 April 2009; pp. 1–29.
4. Hill, C.A.; Such, M.C.; Chen, D.; Gonzalez, J.; Grady, W.M. Battery Energy Storage for Enabling Integration of Distributed Solar Power Generation. *IEEE Trans. Smart Grid.* **2012**, *3*, 850–857.

5. Gong, X.; Huang, Z.; Li, L.; Lu, H.; Liu, S.; Wu, Z. A New State Of Charge Estimating For Lithium Ion Battery Based On Sliding-Mode Observer And Battery Status. In Proceedings of the 35th Chinese Control Conference, Chengdu, China, 27–29 July 2016; pp. 8693–8697.
6. Pham, V.L.; Khan, A.B.; Nguyen, T.T.; Choi, W. A low cost, small ripple and fast balancing circuit for lithium-ion batteries strings. In Proceedings of the IEEE Transportation Electrification Conference and Expo, Asia-Pacific, (ITEC), Busan, Korea, 1–4 June 2016; pp. 861–865.
7. Hemmati, R.; Saboori, H. Emergency of hybrid energy storage systems in renewable energy and transport applications—A review. *Renew. Sustain. Energy Rev.* **2016**, *65*, 11–23.
8. Rodrigues, E.M.; Godina, R.; Osório, G.J.; Lujano-Rojas, J.M.; Matias, J.C.; Catalão, J.P. Assessing Lead-Acid Battery design parameters for energy storage applications on insular grids: A Case Study of Crete and Sao Miguel Islands. In Proceedings of the IEEE International Conference on Computer as Tool (EUROCON), Salamanca, Spain, 8–11 September 2015; pp. 1–6.
9. Venkatesh, J.; Chen, S. Tinnakornsrisuphap, P.; Rosing, T.S. Lifetime-dependent Battery Usage Optimization for Grid-Connected Residential Systems. In Proceedings of the Modelling and Simulation of Cyber-Physical Energy Workshop, Seattle, WA, USA, 13 April 2015; pp 1–6.
10. Serna-Suárez, I.D.; Ordóñez-Plata, G.; Petit-Suárez, J.F.; Caicedo, G.C. Storage systems scheduling effects on the life of lead-acid batteries. In Proceedings of the IEEE PES Innovative Smart Grid Technologies Latin America (ISGT LATAM), Montevideo, Uruguay, 5–7 October 2015; pp. 740–745.
11. Yan, W.; Tian-ming, Y.; Bao-jie, L. Lead-acid Power Battery Management System Based on Kalman Filtering. In Proceedings of the IEEE Vehicle Power and Propulsion (VPPC), Harbin, China, 3–5 September 2008; pp. 1–6.
12. Khayat, N.; Karami, N. Adaptive Techniques Used for Lifetime Estimation of Lithium-Ion Batteries. In Proceedings of the Third International Conference on Electrical, Electronics, Computer Engineering and their Applications (EECEA), Beirut, Lebanon, 21–23 April 2016; pp. 98–103.
13. Khatibi, M.; Jazaeri, M. An analysis for increasing the Penetration of Renewable Energies by Optimal Sizing of Pumped-Storage Power Plants. In Proceedings of the IEEE Electrical Power & Energy Conference, Vancouver, BC, Canada, 6–7 October 2008; pp. 1–5.
14. Seo, H.R.; Kim, G.H.; Kim, S.Y.; Kim, N.; Lee, H.G.; Hwang, C.; Park, M.; Yu, I.K. Power Quality Control Strategy for Grid-connected Renewable Energy Soures Using PV array and Supercapacitor. In Proceedings of the Electrical Machines and Systems (ICEMS) Conference, Incheon, Korea, 10–13 October 2010; pp. 437–441.
15. Delimustafic, D.; Islambegovic, J.; Aksamovic, A.; Masic, S. Model of a hybrid renewable energy system: Control, supervision and energy distribution. In Proceedings of the 2011 IEEE International Symposium Industrial Electronics (ISIE), Gdansk, Poland, 27–30 June 2011; pp. 1081–1086.
16. Belhadji, L.; Bacha, S.; Munteanu, I.; Roye, D. Control of a small variable speed pumped- storage power plant. In Proceedings of the 2013 Fourth International Conference Power Engineering, Energy and Electrical Drives (POWERENG), Istanbul, Turkey, 13–17 May 2013; pp. 787–792.
17. Thounthong, P.; Sikkabut, S.; Sethakul, P.; Davat, B. Control Algorithm of Renewable Energy Power Plant Supplied By Fuel Cell/Solar Cell/ Supercapacitor Power Source. In Proceedings of the International Power Electronics Conference, Sapporo, Japan, 21–24 June 2010; pp. 1155–1162.
18. Farhadi, M.; Mohammed, O. Energy Storage Systems for High Power Applications. In Proceedings of the Industry Applications Society, Annual Meeting, Addison, TX, USA, 18–22 October 2015; pp. 1–7.
19. Alhamad, I.M. A feasibility study of roof mounted grid-connected PV solar system under Abu Dhabi net metering scheme using HOMER. In Proceedings of the Advances in Science and Engineering Technology International Conference, (ASET), Abu Dhabi, United Arab Emirates, 6 February–5 April 2018; pp. 1–4.
20. Pan, C.; Liang, Y.; Chen, L.; Chen, L. Optimal Control for Hybrid Energy Storage Electric Vehicle to Achieve Energy Saving Using Dynamic Programming Approach. *Energies* **2019**, *12*, 588.
21. Werkstetter, S. Existing and Future Ultra-capacitor Applications in the Renewable Energy Market. In Proceedings of the PCIM Europe International Exhibition and Conference for Power Electronics, Intelligent Motion, Renewable Energy and Energy Management, Nuremberg, Germany, 20–22 May 2014; pp. 1–7.
22. Chotia, I.; Chowdhury, S. Battery Storage and Hybrid battery supercapacitor storage systems: A Cooperative critical review. In Proceedings of the Smart Grid Technologies–Asia (ISGT ASIA) Innovative, Bangkok, Thailand, 3–6 November 2015; pp. 1–6.

23. Zhou, H.; Bhattacharya, T.; Tran, D.; Siew, T.S.T.; Khambadkone, A.M. Composite Energy Storage System Involving Battery and Ultra Capacitor with Dynamic Energy Management in Microgrid Applications. *IEEE Trans. Power Electron.* **2011**, *26*, 923–930.
24. Harpool, S.; von Jouanne, A.; Yokochi, A. Supercapacitor Performance Characterization for Renewable Applications. In Proceedings of the IEEE Conference on Technologies for Sustainability, (SusTech), Portland, OR, USA, 24–26 July 2014; pp. 160–164.
25. Keshan, H.; Thornburg, J.; Ustun, T.S. Comparison of Lead-Acid and Lithium Ion Batteries for Stationery Storage in Off-Grid Energy. In Proceedings of the IET Clean Energy and Technology Conference, (CEAT 2016), Kuala Lumpur, Malaysia, 14–15 November 2016.
26. Dutt, D. Life cycle analysis and recycling techniques of batteries used in renewable energy applications. In Proceedings of the International Conference on New Concepts in Smart Cities: Fostering Public and Private Alliances (Smart MILE), Gijon, Spain, 11–13 December 2013; pp. 1–7.
27. Tian, Y.; Li, D.; Tian, J.; Xia, B. A comparative study of state of charge estimation algorithms for lithium ion batteries in wireless charging electric vehicles. In Proceedings of the 2016 IEEE PELS Workshop on Emerging Technologies: Wireless Power Transfer (WoW), Knoxville, TN, USA, 4–6 October 2016; pp. 186–190.
28. Rivera-Barrera, J.; Muñoz-Galeano, N.; Sarmiento-Maldonado, H. Sarmiento-Maldonado. SOC Estimation for Lithium-ion Batteries: Review and Future Challenges. *Electronics* **2017**, *6*, 102.
29. He, H.; Xiong, R.; Fan, J. Evaluation of Lithium-Ion Battery Equivalent Circuit Models for State of Charge Estimation by and Experimental Approach. *Energies* **2011**, *4*, 582–589.
30. Zhang, C.; Jiang, J.; Zhang, L.; Liu, S.; Wang, L.; Loh, P.A. Generalized SOC-OCV Model for Lithium-Ion Batteries and the SOC Estimation for LNMCO Battery. *Energies* **2016**, *9*, 900.
31. Eckhouse, B.; Pogkas, D.; Chediak, M. How Batteries Went from Primitive Power to Global Domination. 13 June 2018. Available online: https://www.bloomberg.com/news/articles/2018-06-13/how-batteries-went-from-primitive-power-to-global-domination (accessed on 20 July 2018).
32. Curry, C. Lithium-ion Battery Costs and Market. 5 July 2017. Available online: https://data.bloomberglp.com/bnef/sites/14/2017/07/BNEF-Lithium-ion-battery-costs-and-market.pdf (accessed on 4 August 2018).
33. Independent, Tesla's Giant Battery Reduces Cost of Power Outages by 90 per cent in South Australia. 12 May 2018. Available online: https://www.independent.co.uk/news/world/australasia/tesla-giant-battery-south-australia-reduce-cost-power-outage-backup-system-fcas-a8348431.html (accessed on 16 July 2018).
34. Asif, A.; Singh, R. Further Cost Reduction of Battery Manufacturing. *Batteries* **2017**, *3*, 17.
35. Nko, M.; Chowdhury, S.P.D. Storage Batteries the Future for Energy Storage. In Proceedings of the 2018 IEEE PES/IAS Power Africa, Cape Town, South Africa, 28–29 June 2018; pp. 705–709.
36. Diouf, B.; Pode, R. Potential of Lithium-Ion Batteries in Renewable Energy. *Renew. Energy* **2015**, *76*, 375–380.
37. Mirsaeidi, S.; Gandomkar, M.; Miveh, M.R.; Gharibdoost, M.R. Power system Load Regulation by Pumped Storage Power Plants. In Proceedings of the 17th Conference on Electrical Power Distribution Networks (EPDC), Tehran, Iran, 2–3 May 2012; pp. 1–5.
38. Banshwar, A.; Sharma, N.K.; Sood, Y.R.; Srivastava, R. Determination of Optimal Capacity of Pumped Storage Plant by Efficient Management of Renewable Energy Sources. In Proceedings of the IEEE Students Conference on Engineering and Systems (SCES), Allahabad, India, 6–8 November 2015; pp. 1–5.
39. Koussa, D.S.; Koussa, M. HOMER Analysis for Integrating Wind Energy into the Grid in Southern of Algeria. In Proceedings of the International Renewable and Sustainable Energy Conference (IRSEC), Ouarzazate, Morocco, 17–19 October 2014; pp. 360–366.
40. Kumari, J.; Subathra, P.; Moses, J.E.; Shruthi, D. Economic analysis of hybrid energy system for rural electrification using HOMER. In Proceedings of the International Conference on Innovations in Electrical, Electronics, Instrumentation and Media Technology, Coimbatore, India, 3–4 February 2017; pp. 151–156.
41. Sureshkumar, U.; Manoharan, P.S.; Ramalakshmi, A.P.S. Economic cost analysis of hybrid renewable energy system using HOMER. In Proceedings of the IEEE-International Conference on Advances in Engineering, Science and Management (ICAESM-2012), Nagapattinam, Tamil Nadu, India, 30–31 March 2012; pp. 94–99.

© 2019 by the authors. Licensee MDPI, Basel, Switzerland. This article is an open access article distributed under the terms and conditions of the Creative Commons Attribution (CC BY) license (http://creativecommons.org/licenses/by/4.0/).

Article

Fault Ride-through Power Electronic Topologies for Hybrid Energy Storage Systems

Ramy Georgious [1,*], Jorge Garcia [1], Mark Sumner [2], Sarah Saeed [1] and Pablo Garcia [1]

1. LEMUR Research Group, Deprtment of Electrical, Electronic, Computers and Systems Engineering, University of Oviedo. 33204 Gijon, Spain; garciajorge@uniovi.es (J.G.); saeedsarah@uniovi.es (S.S.); garciafpablo@uniovi.es (P.G.)
2. PEMC Research Group, Deprtment of Electrical and Electronic Engineering, University of Nottingham, Nottingham NG7 2RD, UK; mark.sumner@nottingham.ac.uk
* Correspondence: georgiousramy@uniovi.es
† This paper is an extended version of our paper published in 2016 IEEE Energy Conversion Congress and Exposition (ECCE), Milwaukee, WI, USA, 18–22 September 2016; pp. 1–8.

Received: 4 December 2019; Accepted: 1 January 2020; Published: 4 January 2020

Abstract: This work presents a fault ride-through control scheme for a non-isolated power topology used in a hybrid energy storage system designed for DC microgrids. The hybrid system is formed by a lithium-ion battery bank and a supercapacitor module, both coordinated to achieve a high-energy and high-power combined storage system. This hybrid system is connected to a DC bus that manages the power flow of the microgrid. The power topology under consideration is based on the buck-boost bidirectional converter, and it is controlled through a bespoke modulation scheme to obtain low losses at nominal operation. The operation of the proposed control scheme during a DC bus short-circuit failure is shown, as well as a modification to the standard control to achieve fault ride-through capability once the fault is over. The proposed control provides a protection to the energy storage systems and the converter itself during the DC bus short-circuit fault. The operation of the converter is developed theoretically, and it has been verified through both simulations and experimental validation on a built prototype.

Keywords: hybrid; energy storage system; buck-boost converter; fault ride-through capability

---

1. Introduction

Power quality is a major concern in modern power systems, particularly in weak microgrids. The concern for the economic importance of power quality issues has led to the development of standards and regulations that define the requirements for equipment and utilities in grid applications [1,2].

Faults in power systems are one of the major causes of power quality issues. Depending on the proximity to the system under consideration, the effects vary substantially. From variations in the voltage and current waveforms parameters (amplitude, frequency and phase) to voltage sags and, in extreme cases, even to voltage outages. Therefore, a continuous research is being done in turning the electric system and its components to fault-tolerant, to boost and develop a more resilient electric grid [3].

In particular, the effects of voltage issues in microgrids have attracted a lot of attention from the research community. One particular case is when the Point of Common Coupling (PCC) is implemented though a Power Electronic Converter (PEC), which interfaces the AC grid with a DC bus. From this DC bus, the microgrid can be supplied in AC, DC or hybrid AC/DC lines [4]. However, the control of the DC bus voltage at the DC side of the grid interfacing converter is critical to ensure adequate operation of the system [5]. In an increasing number of applications, Energy Storage Systems (ESSs) are connected to this DC bus through dedicated PECs, aiming to primarily provide the needed energy in case the microgrid operates in islanding mode. Additionally, the ESS balances the energy flows of the microgrid, accounting for stochastic behavior of distributed generation and loads, therefore decreasing this random factor in the power consumed from the grid [5]. In any case, the sizing of the accumulator is carried out considering energy requirements [6].

Upon changes in the microgrid power flows (due variable load profiles or injection of power from distributed generators), or even voltage variations in the grid, imply transient fluctuations in the DC bus. Usually, the ESSs used for energy supply in microgrids present limited dynamics, and the large transient power spikes reduce their operating lifetime (e.g., in electrochemical batteries) [6]. In these cases, a fast, high-power capability storage system (e.g., Supercapacitor Module (SM)) can be included, forming a Hybrid Energy Storage System (HESS) [6–8]. With an adequate coordinated control, these hybrid systems ensure a stiff behavior of the DC bus, decoupling the grid and the microgrid sides. However, also, they enhance the system reliability by preventing the low-dynamic storage systems to provide large current spikes, resulting in an increase of the system lifetime [9]. HESSs are gaining increased research interest due to their potential benefits in power and energy support in grid applications [5].

Fault-tolerant HESSs are intended to provide these capabilities but at the same time being able to deal with fault conditions in the grid. Depending on the magnitude and the distance to the fault location, the induced variations on the DC bus might vary substantially [4]. For faults far away from the PCC, at distribution or even at transmission levels, the effects within the microgrid are generally limited to voltage sags that can be solved by the regular operation of the HESS. However, for faults at distribution level, closer to the PCC, or even at the DC bus inside the microgrid, more severe voltage variations, and even short-circuit currents might appear through the storage subsystem [4]. Fault-tolerant topologies prevent these dangerous short-circuit currents to circulate across the storage subsystem [10]. In addition, fault ride-through is also expected in this case, therefore once the fault is removed, the system is able to automatically operate properly again in a reasonable amount of time, in accordance to the standards/regulations and the expected behavior of the microgrid [4].

Usually, fault-tolerant topologies make use of additional series switches that interrupt the current flow if a short-circuit appears at the DC bus [4]. In addition to the increase of the system cost, larger losses appear in regular operation due these series switches. This work proposes a non-isolated topology for a fault-tolerant hybrid storage system with fault ride-through capability, suitable for low to medium power applications in microgrids. Its operation is described, and a commutation scheme is proposed to keep the losses in the range of a simple, non-fault-tolerant solution. The performance of the system is validated through both simulations and experimental results on a working prototype of 10 kW.

The simplest PEC topology for interfacing two energy storage devices that build up the HESS to a DC microgrid is the direct connection of two parallel bidirectional boost converters to the DC bus [7,11–18]. This scheme is depicted in Figure 1. This is a cost-effective and reliable solution for low to medium power range applications if galvanic isolation is not required, as the number of elements and devices is relatively low. The case under study considers the simple parallel connection of two distinct energy storage devices. One port consists of a lithium-ion Battery Bank (BB), which will provide a high energy density with slow dynamic response. The other port interfaces to a SM intended to support a high power density and faster dynamic response [5,11–17,19–23]. Therefore, provided that the control strategy is managed correctly, the resulting HESS has a better overall performance as compared to any of the individual systems. This ultimately provides a sustained, high-power, high-dynamic performance of the resulting storage system, also extending the BB lifetime [11,13,21,24].

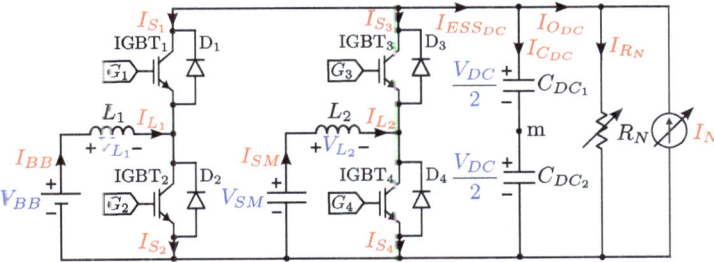

Figure 1. Two parallel bidirectional boost converters connected to BB and SM and sharing the DC bus.

In addition to the lack of galvanic isolation, a major disadvantage of this system is its sensitivity to short-circuit faults at the DC bus. If a short-circuit occurs, the current drawn from both the BB and the SM will increase without control, as the anti-parallel diodes of the upper switches in the legs of the boost converters would allow large short-circuit currents as depicted in Figure 2. This will cause damage to the inductors, the storage devices (BB and SM), and the switches themselves [3,4].

This work describes a new design approach for the power topology and control method which limits operation during DC side short-circuit faults. Even though the number of switching devices in the system is increased in order to build the power topology, the proposed control during the healthy condition achieves an operation without an increase in the losses compared to the two parallel bidirectional boost converters. In addition, the proposed control during the fault condition provides a protection to the converter and the EESs and afford a fast system recovery procedure so as to charge the DC bus once the fault has been cleared.

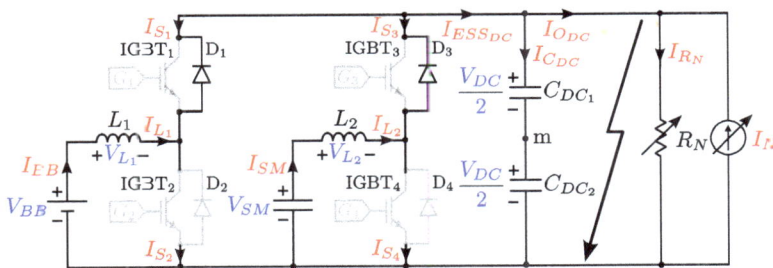

Figure 2. Short-circuit fault at the DC bus of the two bidirectional boost converters.

This paper is organized as follows: Section 2 proposes the buck-boost bidirectional circuit-based topology as a DC short-circuit fault-tolerant topology compared to the simplest boost-based solution shown in Figure 1. The control scheme for the aforementioned boost-based topology is analyzed, in order to establish a starting point in the discussion for the fault-tolerant control adaptation. Then, the control scheme adaptation for the parallel connection buck-boost topologies under consideration is proposed, both in healthy and fault conditions, respectively. After that, Section 3 shows the validation of the proposed system through simulations and experimental results to demonstrate the performance of the proposed solution. In this section, a comparison is also included between the original boost converters and the buck-boost solution. Finally, Section 4 summarizes the work done and discusses future developments.

2. Fault-Tolerant Topologies

The solution to the DC bus fault behavior of the boost-based topology is the connection of a device able to interrupt, or at least limit, the fault currents flowing through the storage units. One option is to connect switches in series with the storage units and the inductors of the converters (see Figure 3).

These switches can be opened during fault conditions to prevent the BB and SM short-circuit currents. In addition, to allow a discharge path for any current flowing through the inductors once the series switches are open, additional free-wheeling switches for each leg are required. Otherwise, a voltage spike will occur, causing arcing or even destruction of the switches. This yields to a final configuration of two parallel bidirectional buck-boost converters [25,26], as shown in Figure 4. This topology has not been analyzed for this particular challenge, i.e., fault tolerance in HESS.

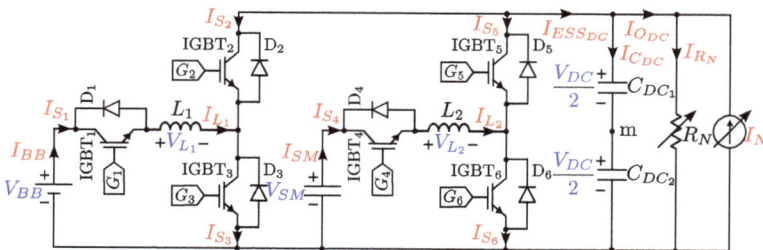

Figure 3. Two parallel bidirectional boost converters with a switch in series between the storage devices and inductors.

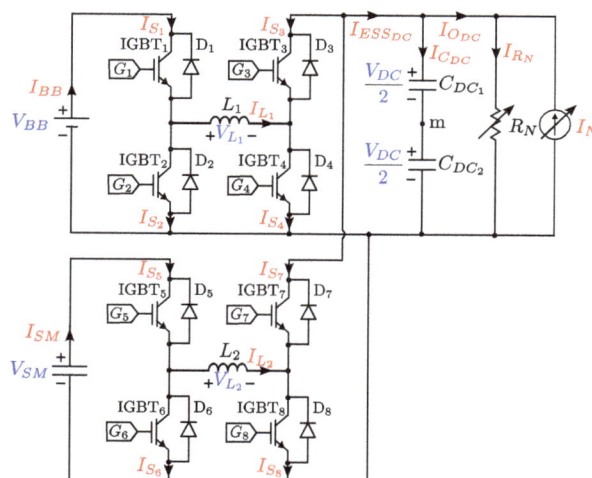

Figure 4. Two parallel bidirectional buck-boost converters connected to BB and SM and sharing the DC bus.

Buck-boost converters can operate either as step-up or step-down voltage interface. In the system under consideration, configured as in Figure 4, the BB voltage (V_{BB}) and the SM voltage (V_{SM}) ratings are less than the DC bus voltage (V_{DC}) and therefore the buck-boost converters will always operate as step-up converter. The main feature discussed in this work is the operation under fault condition of the HESS based on the buck-boost solution. Provided that a suitable control strategy is implemented, the proposed solution enables for a swift system reset once the fault is cleared. A proposal for such a fault ride-through feature will also be demonstrated in the following sections.

It can be seen that the inclusion of the short-circuit fault-tolerant features in the converter adds four more switches compared to the original topology (Figure 1), therefore resulting in higher cost and size than in the initial case. However, as it will be demonstrated in the following sections, by using a proper control, even though the number of switches has increased, the losses of the two topologies can be made very similar.

2.1. Analysis of the Boost Topology Based Control

The case under consideration is a microgrid suitable to operate in islanding mode, with a HESS connected to the DC bus at the PCC, as depicted in Figure 4. In these conditions, a cascaded loop approach is used for the control of the power flows in the system. Both the BB and the SM converters are indeed controlled. The main goal of the control of the BB converter is to maintain the DC bus voltage constant, while the aim of the control of the SM converter is to provide or absorb transient power during load variations. The control strategy is implemented through three control loops: one outer voltage control loop that controls the DC bus voltage, plus two inner current loops in order to control the current flowing through the BB and SM inductors [12,13,16,22,27,28], as shown in Figure 5.

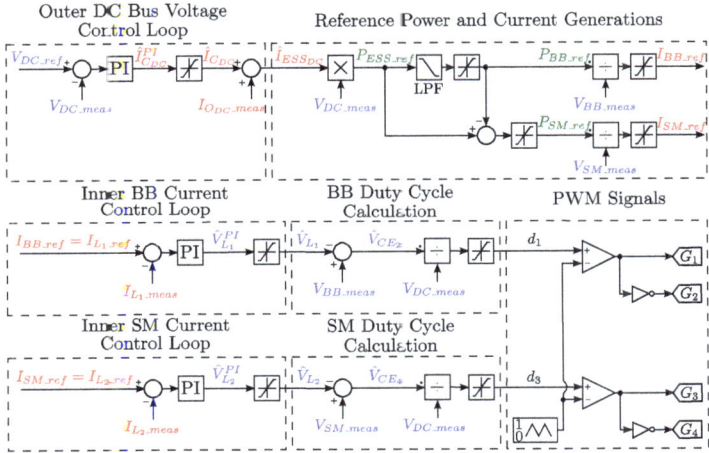

Figure 5. Control of the two parallel bidirectional boost converters to maintain the DC bus voltage constant (BB converter) and provide or absorb transient power during load variations (SM converter).

The outer voltage control loop is a Proportional-Integral (PI) controller to maintain the measured DC bus voltage (V_{DC_meas}) equal to the reference value (V_{DC_ref}). Then, ESS power reference (P_{ESS_ref}) is calculated from the control action of the voltage control loop ($\hat{I}_{C_{DC}}$) and the feed-forward term ($I_{O_{DC\_meas}}$) to improve the recovery of the DC bus due to load variations. The limits of the control action are obtained as follows [29]:

$$I_{C_{DC\_min}} = I_{BB\_min} + I_{SM\_min} + I_{N\_min} - \frac{P_{R_{N\_max}}}{V_{DC\_ref}} \quad (1)$$

$$I_{C_{DC\_max}} = I_{BB\_max} + I_{SM\_max} + I_{N\_max} - \frac{P_{R_{N\_min}}}{V_{DC\_ref}} \quad (2)$$

where:

- $I_{C_{DC\_min}}$ and $I_{C_{DC\_max}}$ are the minimum and maximum current limits of the DC bus current in Amps,
- I_{BB_min} and I_{BB_max} are the minimum and maximum currents of the BB in Amps,
- I_{SM_min} and I_{SM_max} are the minimum and maximum currents of the SM in Amps,
- I_{N_min} and I_{N_min} are the minimum and maximum currents delivered by the rest of the microgrid in Amps,
- V_{DC_ref} is the reference DC bus voltage in Volts,
- $P_{R_{N\_min}}$ and $P_{R_{N\_max}}$ are the minimum and maximum load powers in Watts.

The minimum BB and SM currents are the maximum BB and SM charging currents; however, the maximum BB and SM currents are the maximum BB and SM discharging currents. Then, the SM reference power (P_{SM_ref}) is calculated as the difference between the references for the ESS and BB power values (P_{ESS_ref} and P_{BB_ref}, respectively). A limiter is used to ensure that SM power limits are not exceeded. The BB power reference (P_{BB_ref}) is calculated by using a Low Pass Filter (LPF) to ensure that the SM is providing or absorbing the peak transient power during load variations [28]. Also another limiter is used here, to ensure that the SM provides (or absorbs) the excess power that BB cannot provide (or absorb) during steady state. The power references are calculated according to the following equations [28,30]:

$$P_{ESS\_ref} = (\hat{I}_{C_{DC}} + I_{O_{DC\_meas}}) V_{DC\_meas} \tag{3}$$

$$P_{BB\_ref} = \frac{1}{1 + T_{BB}s} P_{ESS\_ref} \tag{4}$$

$$P_{SM\_ref} = P_{ESS\_ref} - P_{BB\_ref} \tag{5}$$

where:

- P_{ESS_ref}, P_{BB_ref} and P_{SM_ref} are the reference powers of the ESS, BB and SM, respectively, in Watts,
- $\hat{I}_{C_{DC}}$ is the current in the DC bus (control action of the voltage controller) in Amps,
- $I_{O_{DC\_meas}}$ is the measured output current of the two converters in Amps,
- V_{DC_meas} is the measured DC bus voltage in Volts,
- T_{BB} is the time constant of the LPF in Secs,
- s is the Laplace complex variable $s = \sigma + j\omega_d$.

The current references of the BB and SM are obtained by dividing their power references by the corresponding voltage. The bandwidth of the controller for the current in L_2 (inductor in SM converter) (I_{L2}) is faster than the bandwidth of the controller for inductor L_1 (BB converter). This control scheme considers the inductors' voltages, V_{L1} and V_{L2} in Figure 1 to be the control actions at the output of the current regulators. The limits for the inductor voltages can then be calculated as [29]:

$$V_{L1\_min} = V_{BB\_meas} - V_{DC\_ref} \tag{6}$$

$$V_{L1\_max} = V_{BB\_meas} \tag{7}$$

$$V_{L2\_min} = V_{SM\_meas} - V_{DC\_ref} \tag{8}$$

$$V_{L2\_max} = V_{SM\_meas} \tag{9}$$

where:

- V_{L1_min}, V_{L1_max}, V_{L2_min} and V_{L2_max} are the minimum and maximum inductor voltages for the BB and SM boost converters, respectively, in Volts,
- V_{BB_meas} and V_{SM_meas} are the measured storage device voltages in Volts.

Therefore an adaptation between these control actions and the applied duty cycles in both converters, d_1 and d_3, is implemented in the control (Duty Cycle Calculation blocks in Figure 5).

$$d_1 = \frac{V_{BB\_meas} - \hat{V}_{L1}}{V_{DC\_meas}} \tag{10}$$

$$d_3 = \frac{V_{SM\_meas} - \hat{V}_{L2}}{V_{DC\_meas}} \tag{11}$$

where:

- d_1 and d_3 are the duty ratios of the BB and SM converters, respectively,

- $\hat{V}_{L1}$ and $\hat{V}_{L2}$ are the inductor voltages (control action of the current controller) for the BB and SM boost converters, respectively, in Volts.

2.2. Fault-Tolerant Converter Control Strategy for Normal Operation

Some modifications in the previous control are required in order to calculate the duty cycle from the output of the regulator for the buck-boost-based solution (Duty Cycles Calculation block), as shown in Figure 6. With this direct approach, the diagonal switches (S_2 and S_3) and (S_6 and S_7) will commutate with the values of the duty cycles for the BB and SM converters respectively, while the other diagonal switches (S_1 and S_4) and (S_5 and S_8) will switch in a complementary scheme.

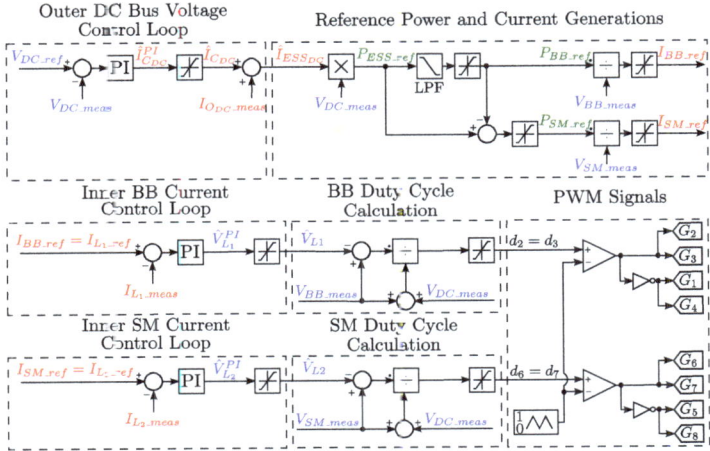

Figure 6. Control of two parallel bidirectional buck-boost converters with the same carrier to maintain the DC bus voltage constant (BB converter) and provide or absorb transient power during load variations (SM converter).

The expressions to calculate the limits for the inductor voltages in this case are as follows:

$$V_{L1\_min} = -V_{DC\_ref} \tag{12}$$

$$V_{L1\_max} = V_{BB\_meas} \tag{13}$$

$$V_{L2\_min} = -V_{DC\_ref} \tag{14}$$

$$V_{L2\_max} = V_{SM\_meas} \tag{15}$$

The same procedure is implemented to calculate the duty ratio for both the BB and SM converters:

$$d_2 = d_3 = \frac{V_{BB\_meas} - \hat{V}_{L1}}{V_{BB\_meas} + V_{DC\_meas}} \tag{16}$$

$$d_6 = d_7 = \frac{V_{SM\_meas} - \hat{V}_{L2}}{V_{SM\_meas} + V_{DC\_meas}} \tag{17}$$

As discussed, all the switches are commutating at High Frequency (HF). This approach, though, will increase the switching losses and therefore the total efficiency of the system will drop. In order to maintain the efficiency, the proposed modifications in the control strategy makes use of two independent modes of operation for each converter during the healthy condition (Normal operation). These modes are Buck mode and Boost mode. This scheme aims to decrease the number of commutating switches in each converter, in order to consequently decrease the switching losses [31–33].

A triangular Pulse-Width Modulation (PWM) technique is implemented to obtain the switching patterns of the switches at the converter. However, in order to achieve a swift transition between the two switching modes, this modulation will be based on two different triangle carrier signals: one carrier for the Boost mode (using peak values of the triangular modulating waveform from 0.0 to 0.5), and another carrier for the Buck mode (using values from 0.5 to 1.0).

This structure implies no overlapping of the HF switching intervals, which yields two different switching patterns for the switches. For example (as shown in Figure 7), if the desired duty cycle is between 0.0 and 0.5, the bidirectional buck-boost converter operates in Boost mode, and therefore switches S_2 and S_6 are turned off, while switches S_1 and S_5 remain turned on continuously. The switches S_3 and S_7 switch with the value of the duty cycle and the switches S_4 and S_8 are their complement. For the Buck mode, when the duty cycle is between 0.5 and 1.0, switches S_3 and S_7 are turned on and switches S_4 and S_8 remain off continuously. S_1 and S_5 switch with value of the duty cycle and S_2 and S_6 are their complement. Figure 8 depicts the implementation of this dual carrier control for the two parallel bidirectional buck-boost converters. The limits of the inductor voltage are the same as in the case of the boost converter (6)–(9).

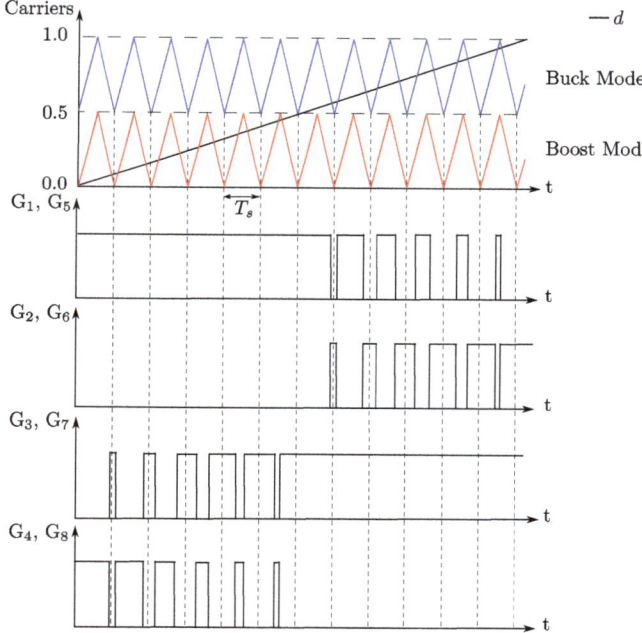

Figure 7. PWM of the two parallel bidirectional buck-boost converters based on two different carriers.

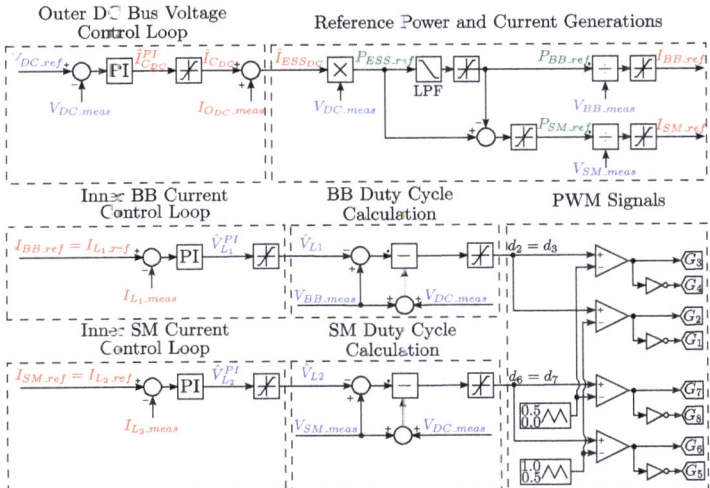

Figure 8. Control of the two parallel bidirectional buck-boost converters with two different carriers to maintain the DC bus voltage constant (BB converter) and provide or absorb transient power during load variations (SM converter).

2.3. Fault-Tolerant Converter Control Strategy under DC-Link Fault Operation

The most critical short-circuit fault types in DC microgrids are either short-circuit between positive and negative bus bars, or a short-circuit between any bus bar and ground [34]. In the first approach of the proposed control scheme, once a DC bus short-circuit fault is detected (for instance by detecting a DC bus voltage below a threshold level), all the switches of the storage converters will be turned off. Therefore the storage devices are instantly disconnected from the DC bus while allowing a discharge path for the inductors at the converters through the anti-parallel diodes of the switches. After the inductances are discharged, no more energy is interchanged between the HESS and the DC bus. However, this control scheme does not have ride-through capability, and therefore even if the fault is removed, the system by itself has no ability of returning to the initial operation mode, unless the control is reset manually and the DC bus is charged externally. After the DC bus is back at rated values, the control scheme works again, and the HESS will remain to support the microgrid normal operation. This scheme is shown in Figure 9.

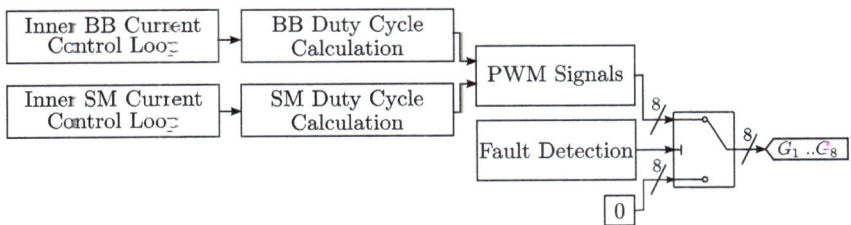

Figure 9. Control scheme including the fault detection block and the pulse disabling.

However, by making a relatively simple modification to the control scheme, the converter can still operate in a controlled manner under fault conditions and can resume normal healthy operation once the fault is over. This modification is introduced in Figure 10.

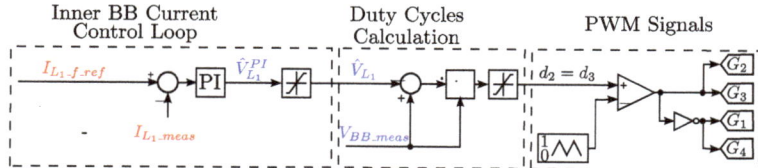

Figure 10. Proposed control of the two parallel bidirectional buck-boost converters during the DC bus fault.

The fault sequence operation of this control scheme is outlined in Figure 11. The DC bus fault occurs at t_f. Then, the fault ride-through capability of the proposed strategy is achieved by providing a small safe current reference ($I_{L_1\_f\_ref}$) value for the BB converter only under DC bus short-circuit. While the DC bus fault is still present, the DC bus voltage remains nearly zero. However, this small current enables the DC bus capacitance to charge linearly once the fault is cleared at t_c. Once a threshold value is reached at t_n, the system gets back to the normal operation scheme, and the standard control takes the system back to the steady state at t_{ss}. This current reference value must be low enough as not to discharge the BB in a reasonable time frame. On the other side, this value must be large enough as to allow a fast detection of the fault clearance condition. Ultimately, this reference value is a function of the DC bus voltage rating and the DC bus capacitance.

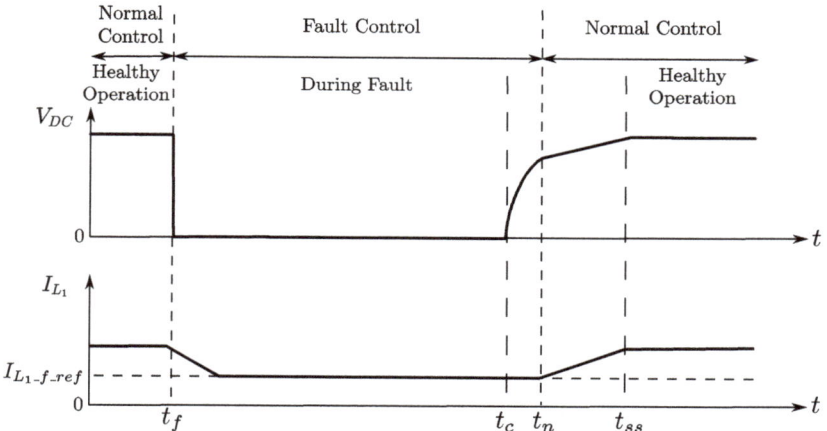

Figure 11. Performance of the fault ride-through control sequence.

The current reference is automatically implemented considering that the voltage across the inductor is limited, and therefore the duty cycle for BB is given by (18) to (20). It also must be noticed that if the fault is permanent, the control is designed to operate for a specific time and then the switches of the BB converters will be turned off in order to decrease the power dissipated from the BB. Another advantage is that this control can be used to charge the DC bus at system start-up. During fault and fault ride-through intervals, the switches of the SM converter are turned off, and SM current is null.

$$V_{L1\_min} = 0 \tag{18}$$

$$V_{L1\_max} = V_{BB\_meas} \tag{19}$$

$$d_2 = d_3 = \frac{V_{BB\_meas} - \hat{V}_{L1}}{V_{BB\_meas}} \tag{20}$$

3. System Parameters and Validation through Simulation and Experimental Results

This section covers the main specifications of the system parameters, the design of the control parameters as well as the validation achieved through simulations of the system performance and a related experimental setup. Simulations of the full system operation have been carried out with MATLAB/SIMULINK/PLECS.

3.1. Main Operating Parameters

Although the conclusions from the prior discussion are valid in general, the validation of such conclusions is demonstrated in a specific laboratory setup that is described in Table 1. The specific control algorithms have been designed and tuned for this particular setup. Once the setup and the control algorithms are defined, then the system performance can be validated through the analysis of both simulations and experimental results.

Table 1. Parameters of the converters.

Parameter	Symbol	Value	Units
Nominal BB voltage	V_{BB}	300	V
Nominal SM voltage	V_{SM}	96	V
Capacitance of the SM	C_{SM}	165	F
DC bus voltage	V_{DC}	500	V
Capacitance of the DC bus	C_{DC}	470	µF
Maximum load power	$P_{F_N\_max}$	2	kW
Inductance of the inductors	L_1 , L_2	21	mH
Resistance of the inductors	R_1 , R_2	0.3	Ω

3.2. Design of the Regulators in the Control Loops

The control scheme has three PI controllers in the outer DC bus voltage control loop and the two inner current control loops for BB and SM. Considering the ideal form of PI controller which is tuned by zero-pole cancellation, the transfer function is given by:

$$C(s) = K_p \left(1 - \frac{1}{sT_i}\right) \quad (21)$$

where:

- $C(s)$ is the transfer function of the PI controller,
- K_p is the proportional gain,
- T_i is the integral time constant.

Current controllers have been tuned by zero-pole cancellation, whereas voltage controller is tuned by loop-shaping techniques.

$$K_{p_{BB}} = 2\pi Bw_{BB} L_1 \quad (22)$$

$$T_{i_{BB}} = \frac{L_1}{R_1} \quad (23)$$

$$K_{p_{SM}} = 2\pi Bw_{SM} L_2 \quad (24)$$

$$T_{i_{SM}} = \frac{L_2}{R_2} \quad (25)$$

The control parameters of the converters are listed in Table 2. Taking into account that the bandwidth of the SM current PI controller (Bw_{SM}) is faster that the bandwidth of the BB current PI controller (Bw_{BB}). Both bandwidths are faster than the bandwidth of voltage PI controller (Bw_v).

Table 2. Parameters of the control of the converters.

Parameter	Symbol	Value	Units
Outer DC Bus Voltage Control Loop			
Bandwidth	Bw_v	30	Hz
Proportional gain	K_{p_V}	0.088548	
Integral time	T_{i_V}	0.141	s
Inner BB Current Control Loop			
Bandwidth	Bw_{BB}	300	Hz
Proportional gain	$K_{p_{BB}}$	39.564	
Integral time	$T_{i_{BB}}$	0.0438	s
Inner SM Current Control Loop			
Bandwidth	Bw_{SM}	500	Hz
Proportional gain	$K_{p_{SM}}$	65.94	
Integral time	$T_{i_{SM}}$	0.0438	s
Cut off frequency of LPF	f_{LPF}	8	Hz
Switching frequency	f_s	20	kHz

3.3. Comparison in Terms of Losses

Special attention has been put on the calculation of the losses in the switches (both conduction and switching losses) during the normal operation of the converters. The conduction and the switching losses of the switches are calculated according to [29]:

$$P_{avg.cond.} = P_{avg.cond.IGBT} + P_{avg.cond.Diode} \tag{26}$$

$$P_{avg.cond.IGBT} = \frac{1}{T_s} \int_0^T (V_{ce}(t) I_c(t)) dt \tag{27}$$

$$P_{avg.cond.Diode} = \frac{1}{T_s} \int_0^T (V_D(t) I_c(t)) dt \tag{28}$$

where:

- $P_{avg.cond.}$, $P_{avg.cond.IGBT}$ and $P_{avg.cond.Diode}$ are the average conduction losses of the switch, IGBT and the anti-parallel diode, respectively, in Watts,
- T_s is the switching time in Secs,
- V_{ce} is the on-state collector emitter voltage of the IGBT in Volts,
- I_c is the on-state collector current of the IGBT in Amps,
- V_D is on-state forward voltage of the anti-parallel diode in Volts.

$$P_{sw} = P_{sw.IGBT} + P_{rec.Diode} \tag{29}$$

$$P_{sw.IGBT} = (E_{on} + E_{off}) f_s \tag{30}$$

$$P_{rec.Diode} = E_{rec} f_s \tag{31}$$

where:

- P_{sw}, $P_{sw.IGBT}$ and $P_{rec.Diode}$ are the switching losses of the switch, the IGBT and the anti-parallel diode, respectively, in Watts,

- E_{on} and E_{off} are the energy loss at IGBT turn on and turn off, respectively, in Joules,
- f_s is the switching frequency in Hz,
- E_{rec} is the energy loss of the reverse recovery of the anti-parallel diode in Joules.

Once these calculations are considered, a comparison between the original boost topology and the new buck-boost topology with both switching schemes (single and dual carrier) has been carried out in PLECS by using the datasheet of the IGBTs (2MBI200HH-120-50 from Fuji Electric) based on the equations explained before. As can be seen from Table 3, the overall losses (both switching and conduction losses) using the fault-tolerant topology with the standard switching pattern are higher than in the original boost-based solution. Still looking at Table 3, it can be noticed how the switching losses at the fault-tolerant topology using the proposed dual carrier control scheme are almost equal to the ones at the original boost-based topology converter. It can also be seen how the conduction losses are higher in the former case, given that S_1 and S_5 are continuously turned on. Therefore, it can be concluded that the total losses at the fault-tolerant solution with the dual carrier scheme are quite similar to the original boost converter case.

Table 3. The losses in the topologies.

Topology	Conduction Losses (W)	Switching Losses (W)	Total losses (W)
Boost	10.67	49.93	60.61
Buck-Boost (original switching mode)	29.7	96	125.7
Buck-Boost (proposed control mode)	22.2	49.93	72.13

Figure 12 shows the operation of both the boost topology and the buck-boost solution with the dual carrier scheme, under healthy conditions. The figure shows that these two solutions give the same performance during transient load steps. The BB controls the DC bus around 500 V, while the SM delivers and absorbs the transient power required during the load steps (from 833.3 W to 1666.7 W and again to 833.3 W) to avoid DC bus voltage variation during the transients. This yields to a fast recovery of the DC bus voltage as well as to a decrease in the power ratings and the stresses (including current ripple) in the battery.

The fault ride-through capability of the buck-boost converter with the proposed dual carrier control is shown in Figure 13. The converters are initially operating under normal control; however, when a DC bus fault is detected at 0.5 s (the DC bus voltage below 15 V threshold), the converters operate under fault control. In fault control mode, a 4 A reference current (value chosen for demonstration purposes) is applied to the BB, while the SM leg is disconnected. When the fault is removed at 2.5 s, this reference will charge the DC bus to a specific value (500 V threshold in this case). Then, the system is automatically reset to the normal control. The DC bus will continue charging linearly until the DC bus reference voltage value and the converter operate in normal mode. The higher current reference, the faster the DC bus charging after fault is removed. This means more unuseful power burned during the fault. At the end, it is a tradeoff depending on the DC bus voltage value, the capacitance value and how fast to charge the DC bus after fault is cleared.

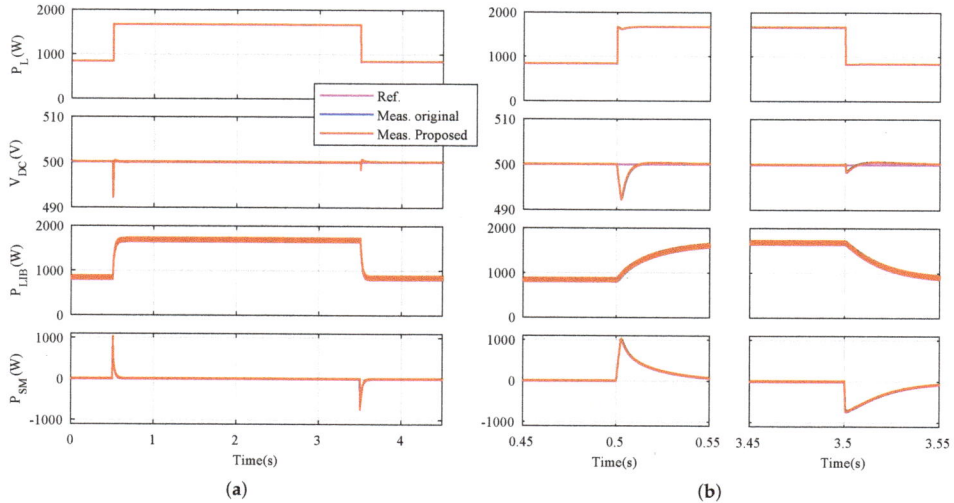

Figure 12. Simulation results during normal operation for the original topology (blue) and the proposed one (red) where: (**a**) the load power (P_L) is changed from 833.3 W to 1666.7 W and then to 833.3 W again. (**b**) zoom at these changes.

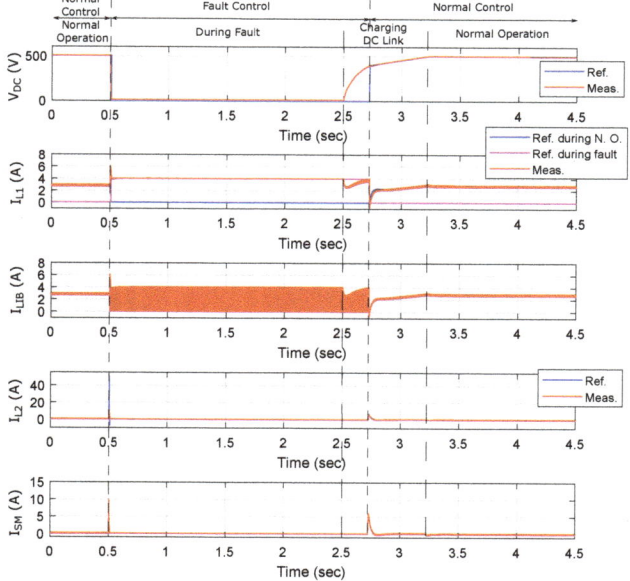

Figure 13. Simulation results during fault and healthy conditions with the proposed control, the fault occurred at 0.5 s and cleared at 2.5 s.

3.4. Experimental Validation of the System Performance

The proposed control scheme in the fault-tolerant topology has been validated using the experimental setup as shown in Figure 14. Figure 15 shows the normal operation of the two parallel bidirectional boost converters and the two parallel bidirectional buck-boost converters. The load is changed at 0 5 s from 833.3 W to 1666.7 W and at 3.6 s is changed again to 833.3W. Figure 15 fully matches with the simulation results in Figure 12. Figure 16 show the operation of the buck-boost converter with the proposed control during the normal operation and fault operation and again fully matches with Figure 13 from simulations. The fault is occurred at 0.5 s and is cleared at 2.4 s.

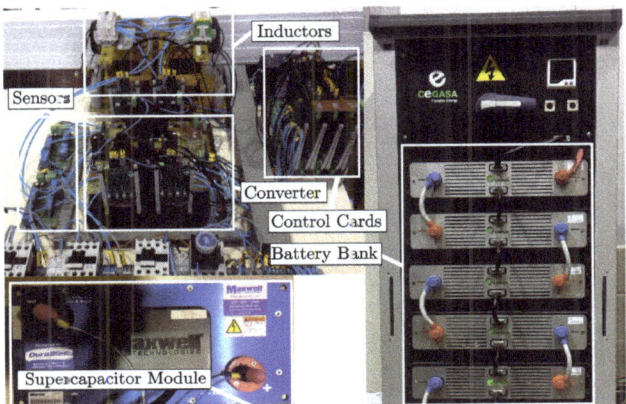

Figure 14. Experimental setup of four legs of IGBTs and can be connected to be boost converter or buck-boost converter.

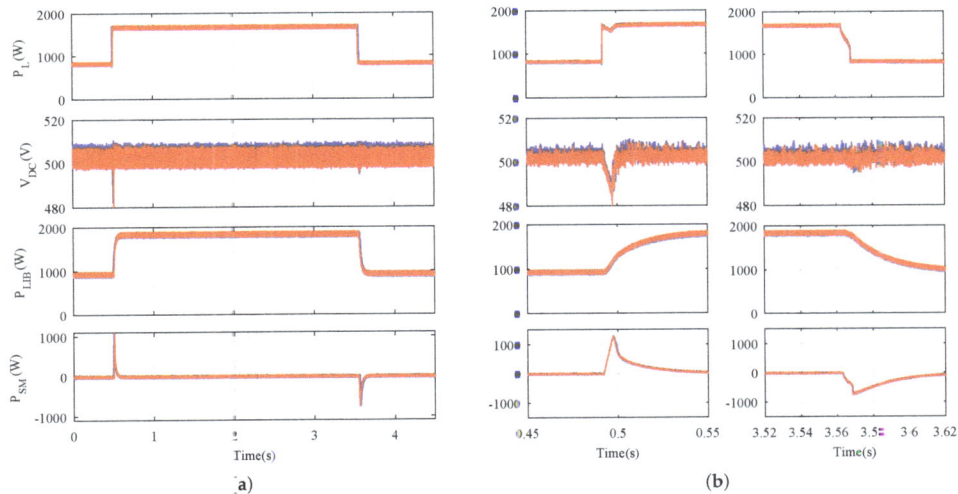

Figure 15. Experimental results during normal operation for the original topology (blue) and the proposed one (red) where: (a) the load power (P_L) is changed from 833.3 W to 1666.7 W and then to 833.3 W again. (b) zoom at these changes.

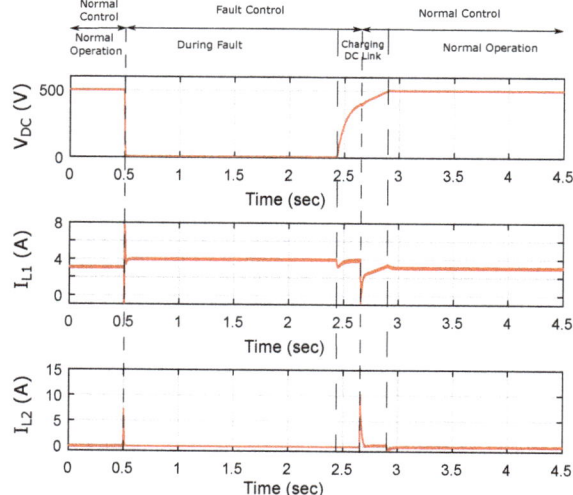

Figure 16. Experimental results during fault and normal operation with the proposed control, the fault occurred at 0.5 s and cleared at 2.4 s.

4. Conclusion and Future Developments

In this paper, a DC bus short-circuit fault-tolerant control scheme for a non-isolated topology for HESSs has been presented, analyzed, verified through simulations and experimentally demonstrated. The proposed control prevents the ESSs and the converter from the damage of the short-circuit current. The proposed strategy includes automatic fault ride-through once the DC bus short-circuit is cleared which helps the converter to operate smoothly in the healthy condition. This configuration has a higher component count than the bidirectional boost version, which is the simplest topology able to achieve the required dynamic performance. However, provided that the proposed dual carrier control scheme during the healthy condition is used, it has been demonstrated that the inclusion of the fault-tolerant, fault ride-through feature does not significantly increase the overall power losses. The proposed configuration, based on the two carrier signals can operate in buck or boost mode, making this scheme also useful for different applications in which the voltage of the ESSs is higher or lower than that of the DC bus.

Future developments include the extension of the study to other fault types, for instance at the storage units; optimization of the control parameters to minimize the energy lost during fault mode; or the extension of this scheme to other kind of applications apart from HESS.

Author Contributions: R.G., J.G. and M.S. conceived the research, designed and performed the experiments; All the authors analyzed the data, and contributed in the discussion and conclusions. All authors have read and agreed to the published version of the manuscript.

Funding: This work has been partially supported by the Spanish Government, Innovation Development and Research Office (MEC), under research grant ENE2016-77919, Project "Conciliator", and by the European Union through ERFD Structural Funds (FEDER). This work has been partially supported by the government of Principality of Asturias, Foundation for the Promotion in Asturias of Applied Scientific Research and Technology (FICYT), under Grant FC-GRUPIN-IDI/2018/000241 and under Severo Ochoa research grants, PA-13-PF-BP13-138 and PF-BP16-133.

Conflicts of Interest: The authors declare no conflict of interest. The founding sponsors had no role in the design of the study; in the collection, analyses, or interpretation of data; in the writing of the manuscript, and in the decision to publish the results.

Abbreviations

The following abbreviations are used in this manuscript:

HESS	Hybrid Energy Storage System
PCC	Point of Common Coupling
PEC	Power Electronic Converter
ESSs	Energy Storage Systems
SM	Supercapacitor Module
BB	Battery Bank
PI	Proportional-Integral
LPF	Low Pass Filter
HF	High Frequency
PWM	Pulse-Width Modulation

The following symbols are used in this manuscript:

V_{BB}	BB voltage	V
V_{SM}	SM voltage	V
V_{DC}	DC bus voltage	V
V_{L1}	Inductor voltages for the BB	V
V_{L2}	Inductor voltages for the SM	V
V_{ce}	On-state collector emitter voltage of the IGBT	V
V_D	On-state forward voltage of the anti-parallel diode	V
$I_{C_{DC}}$	DC bus current	A
$I_{O_{DC}}$	Output current of the two converters	A
I_{BB}	BB current	A
I_{SM}	SM current	A
I_N	Current delivered by the rest of the microgrid	A
I_c	On-state collector current of the IGBT	A
P_{ESS}	ESS power	kW
P_{BB}	BB power	kW
P_{SM}	SM power	kW
P_{R_N}	Load power	kW
$P_{avg.cond.}$	Average conduction losses of the switch	kW
P_{sw}	Switching losses of the switch	kW
d_1	Duty ratio of the BB converter	
d_3	Duty ratio of the SM converter	
C_{SM}	Capacitance of the SM	F
C_{DC}	Capacitance of the DC bus	μF
L_1	Inductance of the inductor connected to the BB	mH
L_2	Inductance of the inductor connected to the SM	mH
R_1	Resistance of the inductor connected to the BB	Ω
R_2	Resistance of the inductor connected to the SM	Ω
$C(s)$	Transfer function of the PI controller	
s	Laplace complex variable	
K_p	Proportional gain	
T_i	Integral time constant	s
T_{BB}	Time constant of the LPF	s
T_s	Switching time	s
f_s	Switching frequency	Hz
Bw	Bandwidth of the PI controller	Hz
E_{on}	Energy loss at IGBT turn on	J
E_{off}	Energy loss at IGBT turn off	J
E_{rec}	Energy loss of the reverse recovery of the anti-parallel diode	J

References

1. Lasseter, R.H. Smart Distribution: Coupled Microgrids. *Proc. IEEE* **2011**, *99*, 1074–1082.[CrossRef]
2. Dugan, R.; McGranaghan, M.; Santoso, S.; Beaty, H. *Electrical Power Systems Quality*; McGraw-Hill Education: New York, NY, USA, 2012.
3. Bui, D.M.; Chen, S.L.; Wu, C.H.; Lien, K.Y.; Huang, C.H.; Jen, K.K. Review on protection coordination strategies and development of an effective protection coordination system for DC microgrid. In Proceedings of the 2014 IEEE PES Asia-Pacific Power and Energy Engineering Conference (APPEEC), Hong Kong, China, 7–10 December 2014; pp. 1–10.[CrossRef]
4. Dragičević, T.; Lu, X.; Vasquez, J.C.; Guerrero, J.M. DC Microgrids—Part II: A Review of Power Architectures, Applications, and Standardization Issues. *IEEE Trans. Power Electron.* **2016**, *31*, 3528–3549.[CrossRef]
5. Tan, X.; Li, Q.; Wang, H. Advances and Trends of Energy Storage Technology in Microgrid. *Int. J. Electr. Power Energy Syst.* **2013**, *44*, 179–191.[CrossRef]
6. Jing, W.; Lai, C.H.; Wong, S.H.W.; Wong, M.L.D. Battery-supercapacitor hybrid energy storage system in standalone DC microgrids: A review. *IET Renew. Power Gener.* **2017**, *11*, 461–469.[CrossRef]
7. Cao, J.; Emadi, A. A New Battery/UltraCapacitor Hybrid Energy Storage System for Electric, Hybrid, and Plug-In Hybrid Electric Vehicles. *IEEE Trans. Power Electron.* **2012**, *27*, 122–132.[CrossRef]
8. Ortuzar, M.; Moreno, J.; Dixon, J. Ultracapacitor-Based Auxiliary Energy System for an Electric Vehicle: Implementation and Evaluation. *IEEE Trans. Ind. Electron.* **2007**, *54*, 2147–2156.[CrossRef]
9. Xiao, J.; Wang, P.; Setyawan, L. Multilevel Energy Management System for Hybridization of Energy Storages in DC Microgrids. *IEEE Trans. Smart Grid* **2016**, *7*, 847–856.[CrossRef]
10. Maharjan, L.; Yamagishi, T.; Akagi, H.; Asakura, J. Fault-Tolerant Operation of a Battery-Energy-Storage System Based on a Multilevel Cascade PWM Converter With Star Configuration. *IEEE Trans. Power Electron.* **2010**, *25*, 2386–2396.[CrossRef]
11. Tummuru, N.R.; Mishra, M.K.; Srinivas, S. Dynamic Energy Management of Hybrid Energy Storage System With High-Gain PV Converter. *IEEE Trans. Energy Convers.* **2015**, *30*, 150–160.[CrossRef]
12. Kollimalla, S.K.; Mishra, M.K.; Narasamma, N.L. Design and Analysis of Novel Control Strategy for Battery and Supercapacitor Storage System. *IEEE Trans. Sustain. Energy* **2014**, *5*, 1137–1144.[CrossRef]
13. Kollimalla, S.K.; Mishra, M.K.; N, L.N. Coordinated Control and Energy Management of Hybrid Energy Storage System in PV System. In Proceedings of the 2014 International Conference on Computation of Power, Energy, Information and Communication (ICCPEIC), Chennai, India, 16–17 April 2014; pp. 363–368. [CrossRef]
14. Sathishkumar, R.; Kollimalla, S.K.; Mishra, M.K. Dynamic Energy Management of Micro Grids Using Battery Super Capacitor Combined Storage. In Proceedings of the 2012 Annual IEEE India Conference (INDICON), Kochi, India, 7–9 December 2012; pp. 1078–1083. [CrossRef]
15. Jayasinghe, S.D.G.; Vilathgamuwa, D.M.; Madawala, U.K. A Direct Integration Scheme for Battery-Supercapacitor Hybrid Energy Storage Systems with the Use of Grid Side Inverter. In Proceedings of the 2011 Twenty-Sixth Annual IEEE Applied Power Electronics Conference and Exposition (APEC), Fort Worth, TX, USA, 6–11 March 2011; pp. 1388–1393. [CrossRef]
16. Li, W.; Joos, G.; Belanger, J. Real-Time Simulation of a Wind Turbine Generator Coupled with a Battery Supercapacitor Energy Storage System. *IEEE Trans. Ind. Electron.* **2010**, *57*, 1137–1145.[CrossRef]
17. Yoo, H.; Sul, S.K.; Park, Y.; Jeong, J. System Integration and Power-Flow Management for a Series Hybrid Electric Vehicle Using Supercapacitors and Batteries. *IEEE Trans. Ind. Appl.* **2008**, *44*, 108–114.[CrossRef]
18. Vazquez, S.; Lukic, S.M.; Galvan, E.; Franquelo, L.G.; Carrasco, J.M. Energy Storage Systems for Transport and Grid Applications. *IEEE Trans. Ind. Electron.* **2010**, *57*, 3881–3895.[CrossRef]
19. Li, W.; Joos, G. A Power Electronic Interface for a Battery Supercapacitor Hybrid Energy Storage System for Wind Applications. In Proceedings of the 2008 IEEE Power Electronics Specialists Conference (PESC), Rhodes, Greece, 15–19 June 2008; pp. 1762–1768. [CrossRef]
20. Khaligh, A.; Li, Z. Battery, Ultracapacitor, Fuel Cell, and Hybrid Energy Storage Systems for Electric, Hybrid Electric, Fuel Cell, and Plug-In Hybrid Electric Vehicles: State of the Art. *IEEE Trans. Veh. Technol.* **2010**, *59*, 2806–2814. [CrossRef]
21. Solero, L.; Lidozzi, A.; Pomilio, J.A. Design of Multiple-Input Power Converter for Hybrid Vehicles. *IEEE Trans. Power Electron.* **2005**, *20*, 1007–1016.[CrossRef]

22. Mendis, N.; Muttaqi, K.M.; Perera, S. Active Power Management of a Super Capacitor-Battery Hybrid Energy Storage System for Standalone Operation of DFIG based Wind Turbines. In Proceedings of the 2012 IEEE Industry Applications Society Annual Meeting (IAS), Las Vegas, NV, USA, 7–11 October 2012; pp. 1–8. [CrossRef]
23. Gee, A.M.; Robinson, F.V.P.; Dunn, R.W. Analysis of Battery Lifetime Extension in a Small-Scale Wind-Energy System Using Supercapacitors. *IEEE Trans. Energy Convers.* **2013**, *28*, 24–33.[CrossRef]
24. van Voorden, A.M.; Elizondo, L.M.R.; Paap, G.C.; Verboomen, J.; van der Sluis, L. The Application of Super Capacitors to relieve Battery-storage systems in Autonomous Renewable Energy Systems. In Proceedings of the 2007 IEEE Lausanne Power Tech, Lausanne, Switzerland, 1–5 July 2007; pp. 479–484.[CrossRef]
25. Tytelmaier, K.; Husev, O.; Veligorskyi, O.; Yershov, R. A Review of Non-Isolated Bidirectional DC-DC Converters for Energy Storage Systems. In Proceedings of the 2016 II International Young Scientists Forum on Applied Physics and Engineering (YSF), Kharkiv, Ukraine, 10–14 October 2016; pp. 22–28. [CrossRef]
26. Du, Y.; Zhou, X.; Bai, S.; Lukic, S.; Huang, A. Review of Non-isolated Bi-directional DC-DC Converters for Plug-in Hybrid Electric Vehicle Charge Station Application at Municipal Parking Decks. In Proceedings of the 2010 Twenty-Fifth Annual IEEE Applied Power Electronics Conference and Exposition (APEC), Palm Springs, CA, USA, 21–25 February 2010; pp. 1145–1151. [CrossRef]
27. Silva, M.A.; de Melo, H.N.; Trovao, J.P.; Pereirinha, P.G.; Jorge, H.M. Hybrid Topologies Comparison for Electric Vehicles with Multiple Energy Storage Systems. In Proceedings of the 2013 World Electric Vehicle Symposium and Exhibition (EVS27), Barcelona, Spain, 17–20 November 2013; pp. 1–8. [CrossRef]
28. Georgious, R.; Garcia, J.; García, P.; Sumner, M. Analysis of hybrid energy storage systems with DC link fault ride-through capability. In Proceedings of the 2016 IEEE Energy Conversion Congress and Exposition (ECCE), Milwaukee, WI, USA, 18–22 September 2016; pp. 1–8. [CrossRef]
29. Georgious, R. A Hybrid Solution for Distributed Energy Storage for Microgeneration in Microgrids: Design of the Electronic Power and Control System. Ph.D. Thesis, University of Oviedo, Oviedo, Spain, 2018.
30. Wang, C.S.; Li, W.; Wang, Y.F.; Han, F.Q.; Meng, Z.; Li, G.D. An Isolated Three-Port Bidirectional DC-DC Converter with Enlarged ZVS Region for HESS Applications in DC Microgrids. *Energies* **2017**, *10*, 446.[CrossRef]
31. Lee, Y.J.; Khaligh, A.; Emadi, A. A Compensation Technique for Smooth Transitions in a Noninverting Buck–Boost Converter. *IEEE Trans. Power Electron.* **2009**, *24*, 1002–1015.[CrossRef]
32. Aharon, I.; Kuperman, A.; Shmilovitz, D. Analysis of Dual-Carrier Modulator for Bidirectional Noninverting Buck–Boost Converter. *IEEE Trans. Power Electron.* **2015**, *30*, 840–848.[CrossRef]
33. Wei, C.L.; Chen, C.H.; Wu, K.C.; Ko, I.T. Design of an Average-Current-Mode Noninverting Buck–Boost DC–DC Converter With Reduced Switching and Conduction Losses. *IEEE Trans. Power Electron.* **2012**, *27*, 4934–4943.[CrossRef]
34. Salomonsson, D.; Soder, L.; Sannino, A. Protection of Low-Voltage DC Microgrids. *IEEE Trans. Power Deliv.* **2009**, *24*, 1045–1053. [CrossRef]

© 2020 by the authors. Licensee MDPI, Basel, Switzerland. This article is an open access article distributed under the terms and conditions of the Creative Commons Attribution (CC BY) license (http://creativecommons.org/licenses/by/4.0/).

MDPI
St. Alban-Anlage 66
4052 Basel
Switzerland
Tel. +41 61 683 77 34
Fax +41 61 302 89 18
www.mdpi.com

Energies Editorial Office
E-mail: energies@mdpi.com
www.mdpi.com/journal/energies

www.ingramcontent.com/pod-product-compliance
Lightning Source LLC
LaVergne TN
LVHW070647100526
838202LV00013B/903